Acclaim for Newt Gingrich and WINDOW OF OPPORTUNITY

"One may violently differ with many of Congressman Gingrich's views while still admiring his intellect, originality, and future consciousness. At a time when both the Democrats and the Republicans are parties of nostalgia, this book can help stir an overdue debate about the issues of tomorrow."

—Alvin Toffler

"A really remarkable intellectual *tour de force*. . . . the first time I have seen a credible alternative to the current 'conventional' wisdom about how the political forces in this country can be mobilized to lead us into a brighter future."

—Hans Mark
Deputy Administrator, NASA

"[WINDOW OF OPPORTUNITY] has the potential of becoming one of the most important political statements of this century."

—Adam Meyerson
Editor, *Policy Review*

"Gingrich gushes with ideas: government bonuses to poor children who learn to read; job training classes and child care centers for people who want to work; tax credits for working class families who buy home computers."

—Steven J. Roberts,
The New York Times

"Growth, opportunity, jobs and progress are the issues of the future. Newt Gingrich's book shows us the way."

—Jack Kemp, Member of Congress

"[Rep. Gingrich is] a practical man of ideas, a deeply committed conservative with a bold vision of the future, a finely honed ability to communicate and an unusual grasp of political strategy."

—Steven K. Beckner,
Conservative Digest

"[Gingrich is] a strategic thinker of the first rank."

—William A. Rusher
Publisher, *National Review*

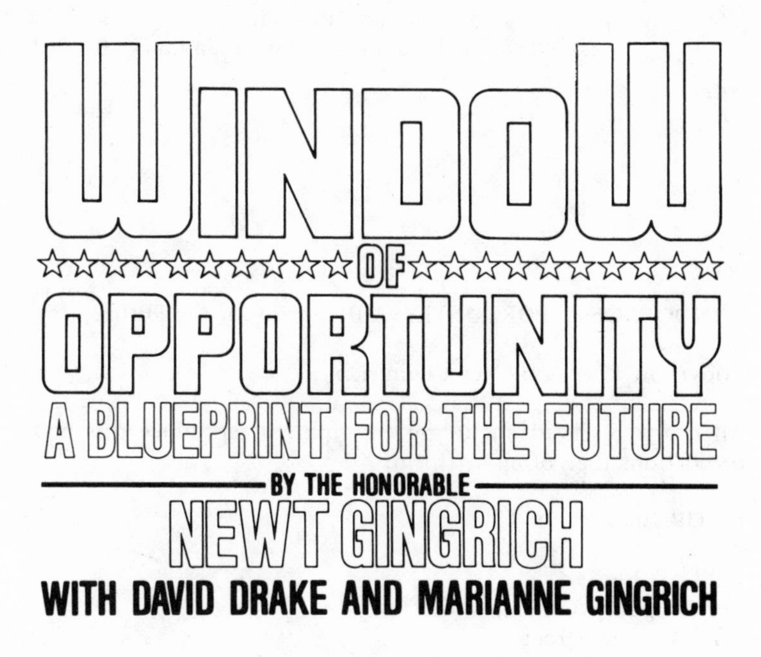

WITH DAVID DRAKE AND MARIANNE GINGRICH

TOR

A TOM DOHERTY ASSOCIATES BOOK

Published in Association with Baen Enterprises

"We have every right to dream heroic dreams."
—President Ronald Reagan
From his inaugural address, January 20, 1981.

WINDOW OF OPPORTUNITY: A Blueprint for the Future

A TOR Book

Published by:
Tom Doherty Associates, Inc.
8-10 West 36 Street
New York, N.Y. 10018

In association with Baen Enterprises, Inc.

ISBN: 0-312-93922-1

Cover art: Alan Gutierrez
Cover design: Carol Russo

Printed in the United States of America

Distributed by:
St. Martin's Press
175 Fifth Avenue
New York, N.Y. 10010

10 9 8 7 6 5 4 3 2 1

ACKNOWLEDGEMENTS

Four people deserve particular thanks for making *Window of Opportunity* a reality. Jerry Pournelle introduced me to Jim Baen and convinced me that writing a book would be an effective way to publicize the ideas of the Conservative Opportunity Society. Jim Baen had the publishing vision to propose a book project to Marianne and me, and to match us up with two very fine professional writers: our coauthor, David Drake, and Janet Morris. Money alone could not buy the creativity, skill, and effort that Janet contributed to the final draft. David's contribution, of course, cannot be overstated.

Jim also had the fortitude to first agree to a backbreaking, well-nigh impossible publication schedule, and then to work with us and hold our hands through the many months of confusion, frustration, and, at times, irritation which it took two busy beginners and two harried professionals to produce the book you are now holding. I enthusiastically recommend him to anyone as a publisher, but particularly to new authors who are more certain of what they have to say than how to say it. Jim has been a remarkable associate and admirable friend.

ACKNOWLEDGMENTS

TABLE OF CONTENTS

PREFACE
Jerry Pournelle

I first heard of Georgia Congressman Newt Gingrich when my telephone rang. Newt had read my book *A Step Farther Out* and wanted to discuss the ideas in it. I have, in the last half century, known several members of Congress, but I have never had that experience before. Congress people are often concerned with constituent matters, and the better ones are very knowledgeable about national affairs handled by committees they're on, but not much else. Congress as an institution has largely forfeited the respect of the people of this nation, in large part precisely because most members of Congress can't be bothered with books about the future. They're too busy running for office.

In a sense it isn't their fault, of course. The American political system makes it highly desirable to be a member of Congress. The pay is not especially high but the influence is enormous—particularly if you're a partner in a law firm. However, the American electoral system insures that nearly all elected officials will spend so much time *getting* their jobs that they will have none left actually to *do* them.

There are notable exceptions. Newt is one of them. If Congress is ever to win back the respect and affection it held in the days of Madison, Jay, Webster, Calhoun, Clay, Douglas, and Lincoln, it will be through the efforts and examples of the tiny minority of members like Gingrich. Far from being a partner in a powerful law firm, Newt is an historian; one of the very few ever to sit in the Congress. He's not ashamed to be an intellectual.

He's also not ashamed to call himself a politician:

a word that's not much in favor these days. I remember that I was shocked the first time he said it of himself. We were sitting in the somewhat shabby opulence the Speaker metes out to junior members of the minority party, discussing the future of the space program.

"The space program is always in trouble," Newt said, "because it has never been championed by a politician. I want to be that politician. You technical people show me what we can do, and I'll try to take it from there."

A few weeks later, in partnership with Congressman Akaka of Hawaii, Newt Gingrich organized the Space Caucus within the House of Representatives. For that alone Newt deserves a place in the history books. I don't know the names of the members of the *Cortes* who advised Isabella to finance Columbus, but perhaps this generation can do better in acknowledging its farsighted statesmen.

The press usually describes Newt Gingrich as "a conservative Republican from Georgia." That's not an unfair description, but it's incomplete. Although his views have won him considerable influence in both Conservative and Republican circles, he is no ideologue. He has plenty to say that Liberals and Conservatives alike will find appealing, because he looks behind slogans and rhetoric of both Liberals and Conservatives to examine the results their policies obtain. America is not, after all, deeply divided in intentions. We all want peace, freedom, and prosperity. The division comes when we translate these high goals into policies and laws.

There was a time when the United States had so much wealth that we could afford drastic errors. That time has long past, and now we're in big trouble. The future is mortgaged. The similarities between the United States today and the last days of the Roman Republic are too great to be ignored, and the end of the road that Rome took is frighteningly clear.

If things are left to go as they have been, this nation will inevitably continue on what Friedrich

Hayek called *The Road to Serfdom*. The result will be a dictatorship gratefully turned to by people exhausted from social conflict and class war. This is not merely possible, but well-nigh inevitable.

Things need not be left to themselves. There is great strength in this nation. We still have the capacity for a bloodless and non-violent revolution. It is not too late for recovery. The repeated shocks of new technologies, new inventions, can overcome even the greatest social inertia—provided we act with skill and intelligence.

This book shows, readably and with clarity, what skillful and intelligent policies can accomplish. The programs described here have been drawn from a hundred sources ranging all across the political spectrum: but they are not a mere hodgepodge of "pragmatic" notions. What is presented are programs united by firm commitment to the principles that built this Republic, stated by an historian who understands what the delegates to the Constitutional Convention intended.

We are only a few years short of the 200th anniversary of that Convention. It is time we recovered the wisdom of the Framers, and applied it to a world so changed by technology that the delegates to the Philadelphia Convention would scarcely recognize it: for although the world changes swiftly, human nature does not. We have mostly the same needs and motivations today as our ancestors had in the years after the Revolution.

America is often described as "pragmatic," meaning that we are guided by "what works" rather than by adherence to some dead ideology. That may have once been true; but today, "pragmatic" generally refers to ways that politicians and bureaucrats can mortgage the future in order to retain office. America needs no ideology—but we must have principles, and indeed the people of this nation always have had. Samuel Johnson once said that we seldom need education, but we often need to be reminded. This book reminds us of our twin heritages: firm principles and large doses of common sense.

* * *

Gingrich spares no sacred cows. He's for a strong America, but that hardly means blind support of the Pentagon. His speeches on military policy are among the most thought-provoking ever given in the Congress, but they've earned him the undying hatred of a lot of the military establishment.

Serving officers in the combat arms have a different view. He's plenty popular with them, because he understands why a republic needs fighting services.

Gingrich is one of the very few members of Congress to have read not only Col. Harry Summers's modern classic *On Strategy*, but also Clausewitz and the older classics from which Summers drew his inspiration; but Newt's study of the military doesn't stop with books. He also spends time with troops in the field, particularly the Georgia national guard.

He can be pretty rough on his friends when they need it. Gingrich was the keynote speaker at the second annual meeting of the L-5 Society's Space Development Conference. He nearly flayed us alive, pointing out in well-informed detail just what kind of mistakes the organization was making in its drive to build up the space budget. "Dairy farmers are better organized than you are," he said. "Which is why they get more money than NASA does."

He's been helping us organize ever since; and each time I talk with him, I'm amazed at how many of the technical details of space exploration he's absorbed.

Military policy; space policy; welfare, medicare, taxes, social security; darned little escapes his microscopic attention. The result is one of the most comprehensive examinations of U.S. policy I've ever had the pleasure to read.

In the late 60's and through most of the 70's the official position of most intellectuals was one of gloom and despair. From California Governor Jerry Brown's "Era of Limits" to President Carter's "national malaise," the official word was that the party was over. Fortune and plenty and growth were not merely impossible, but evil, ambitions. The human race would probably not survive, and if we managed to squeak through it would be without glory. The very few of us who dared say otherwise were considered freaks, and on the few occasions on which we were permitted to address university students we could be certain that a large group of faculty members would be present to assure the students that we were nuts.

That was not a pretty era. We live with its effects to this day. Students were brought up in an atmosphere of gloom; and students are not stupid. It took no great talent in logic to draw the proper conclusions. If there was to be no future, then there was no point to saving. Why deny oneself immediate gratification? Surely it was foolish to save in aid of the next generation, since there would probably be no next generation. Thus was born the "me generation." It was not an effect that the Club of Rome had intended, but it was inevitable given the intellectual climate. If no one will live until the next harvest, there is no point in planting; we might as well eat the seed corn.

Fortunately, university professors do not rule America. William F. Buckley once said that he would rather be governed by the first five hundred names in the Boston telephone book than by the entire faculty of Harvard University, and of course he is right. The stories about "ivory tower" professors have the ring of truth. The one professor to become President of these United States was Woodrow Wilson, whose blind adherence to abstract theories certainly helped prolong World War I, and did little to insure that the Treaty of Versailles would bring lasting peace to Europe or the world.

Once again, Newt Gingrich is a notable exception. He can hold his own with the academicians, for he was once one of them; but unlike Wilson, who sought to teach Washington the lessons of the academy, Newt Gingrich has learned the lessons of the politician—without losing his appreciation for the world of ideas.

During the "era of limits" a few of us dared say that we could not only survive, but survive with style; that humanity has choices, and among them is a vast and glorious future. The only limit to human expansion is nerve; for man does not live on a limited planet. We live in a solar system of nine planets, nearly forty moons, a million asteroids, a billion comets, and a friendly star that puts forth nearly limitless energy for the taking. We can look forward down endless corridors of time to a hundred-billion-year future; we need only grasp the opportunity.

Those were the concepts that Newt Gingrich called to discuss with me. The discussion was not idle chatter. For all his intellectual abilities, Newt Gingrich is not in love with ideas for their own sake; he wants to put them to use, to bring about the glorious future that some dreamers and theorists see. Ideas are fine, but laws and institutions and organizations are their concrete embodiment, and it is these that Newt Gingrich examines in this book.

This is a work born from hundreds of speeches and discussions, from books and newspapers, from student bull sessions and Georgia town meetings, from Universities and cotton farms, from meetings of learned societies and conversations with constituents. It is a remarkable book, almost unique in that, without the slightest compromise with the principles that made this nation great, Gingrich presents a detailed blueprint, a practical program that not only proves that we can all get rich, but shows how.

Today it is nearly impossible to publish a book so bad that some reviewer will not call it "important." We have cheapened the language until only superlatives attract attention, and they not often. In this era of superhype it is difficult to call attention to the truly important. So be it. I can only say that this *is* an important book, one which has the potential to bring real changes to this world; but it is important without being stilted, indeed important because it is readable and entertaining; important because it shows, with clarity and detail, paths we can and must take if we are truly to leave behind the "era of limits" and grasp the limitless opportunities available.

It's raining soup, and Newt Gingrich has the blueprints for soup bowls.

INTRODUCTION

We stand at a crossroads between two diverse futures.

On the one hand, we face a window of vulnerability—a gap between Soviet and American military power. Many analysts feel that we are entering a period during which the Soviet Union will have enough power to pursue adventurist policies around the world without fear of American response. Our window of vulnerability will widen in coming years to include increasing terrorism, escalating conflicts in the Third World, and the inevitable proliferation of chemical, biological and nuclear weapons. All these developments increase our peril.

But while we must be aware of our vulnerabilities and take appropriate steps to survive, no narrowly defensive approach is going to be adequate to our future need: we cannot survive in a "fortress America" or a "fortress free world." America cannot lead the human race into a peaceful, prosperous, free future by focusing narrowly on the threatening aspects of life.

On the other hand, there is a beckoning window of opportunity fully as real as the window of vulnerability. Breakthroughs in computers, biology, and space make possible new jobs, new opportunities, and new hope on a scale unimagined since Christopher Columbus discovered a new world.

Everywhere I turn I see great possibilities for a better world for all mankind, yet politicians, news reporters, and the liberal elite publicly despair of the future.

This book has been written to send a simple

message: there is still hope if we will but seize the opportunity.

There is hope for a continuing revolution in biology which will allow us to feed the entire planet; hope for jobs, opportunities, and adventures in space; hope that computers and information science will allow us to work at home, to help all Americans live fuller lives, to save energy and resources by expanding our potential while exchanging transportation for communication.

We have the potential to enter an era which brings power to the people in a practical and exciting manner. This book is meant to illuminate the great opportunities which beckon America toward an age of exploration and innovation rivaling anything we have known—opportunities that will not last forever.

As a free people we must choose our future and make our choice reality. We have no guarantee that the window of opportunity will lead us into a promised land, but we know that, in Jefferson's words, we are the last best hope of mankind. We must be willing to lead, because no one else will.

Our generation of Americans must decide whether to lead mankind into freedom, productivity, and peace or whether we will preside over the slow decay of mankind into a world of terrorism and tyranny.

I hope to convince you that the world of hope and opportunity is worth working for, and that the difference for your children and your grandchildren between the windows of opportunity and vulnerability is such that we must do all we can to lead mankind to opportunity.

CHAPTER 1: THE WINDOW OF OPPORTUNITY

There exists for the United States today a window of opportunity through which we can look, and—with luck and hard work—reach to create a bright and optimistic future for our children and grandchildren. If, however, we continue the policies of the last twenty years, that window will close and we will bequeath to them a pessimistic future of economic and social decay. Furthermore, if we are both unwise and unlucky, we will deliver our children over to a world of terrorism and nuclear war—no future at all.

At the moment, this window of opportunity is open and the optimistic future is reachable, but reaching it will require changes in our current behavior and institutions. This optimistic future will necessitate accepting the possibilities inherent in our emerging technologies and accelerating the transition to a high technology, information-based society.

This optimistic future will require reforms—both in theory and practice—of the welfare system. The psychology and structure of today's welfare state threatens to cut off the poor from the technological opportunities offered by information processing and computer programming. We want all our children to have an equal chance to participate in the "opportunity society" this book describes.

An optimistic future will require fundamental reforms of government so that it can encourage and help the rest of society make the transition to this new world. Today, government bureaucracies are among the most labor-intensive and technologically backward offices in America. If you have a Merrill Lynch money market account, you can call a toll-free number, punch in your access code, and a computer will tell you your account's balance and the interest rate paid for the previous week and month. But if you try to get information about the taxes you've paid into the Social Security system, you must write a letter and wait months to have the Social Security Administration tell you they are several years behind in processing such information.

The time differentials between these two systems exist not because the Merrill Lynch people are dramatically brighter than the Social Security workers, but because the information-management techniques used by the Social Security Administration are at best obsolete and at worst archaic.

It is the system of high technology and positive management which makes individuals at Merrill Lynch effective, as it is the system of old technology and obsolete public administration techniques which makes the Social Security system relatively non-productive.

We must bring government bureaucracies into the information age. Today we have a horse-and-buggy style of public administration presiding over a nation entering the space-shuttle age. Little wonder, then, that there is growing public dissatisfaction with Washington in particular and with government in general.

A look through the window of opportunity at our optimistic future shows us that we must teach our citizens the fundamentals today that they will need to be effective tomorrow. Alvin Toffler suggests, in his book *The Third Wave*, that an information society will require "Triliteracy," a combination of: 1) the old three R's of reading, writing, and arithmetic; 2) the use of mass media; and 3) the use of computers. When the citizens of our emerging information society have mastered Triliteracy, there will there be a tremendous increase not only in America's productivity, but also in each individual's creativity as people learn to retrain themselves in order to find their own paths to wisdom and knowledge.

This new approach to knowledge-acquisition will also require us to recognize that learning is a lifetime process. In the future, much more of the burden will be on the student than on the teacher. Education will begin with the fundamentals of Triliteracy and that knowledge will help students work their way through a world which will be changing rapidly as long as they live.

The assumption that we can reach the optimistic future we see through our window of opportunity is based on another, more crucial assumption: that the current resurgence of traditional values will continue. It is impossible to envision an optimistic future dominated by drugs, muggings, rapes, murders, and general corruption.

Safety breeds optimism. People must be comfortable in their neighborhoods, able to walk the streets at night, and secure in the knowledge that their children are more likely to learn about computers than cocaine in our public schools. A truly honest society must be governed by ethics rather than fear of punishment, but those ethics must be translated into tougher laws against crime and corruption.

Only by a return to individual honesty and safety can an optimistic future come into being and our children enjoy a world as dramatically superior to ours in personal freedom and opportunity as ours is superior to our great grandparents'.

Think back a little over one hundred years to the period before electricity, chemistry, and the internal-combustion engine revolutionized our world. In 1870, there were none of the modern conveniences we take for granted today—no telephones, vacuum cleaners, dishwashers, or refrigerators. When people were sick, they lived or died without antibiotics. Malaria, yellow fever, typhus, and cholera were deadly facts of life to be endured, not controlled. People traveled on horseback or in wagons, except in the few places where there were railroads. In the world of

1870, entertainment did not include record players, motion pictures, radio, or television.

The people of 1870 would have had a hard time understanding the possibilities of our modern era, yet a great many of them believed in progress: our pioneering forefathers were convinced that mankind would keep on inventing and developing new ideas and that the world would keep improving.

Now we are standing on the verge of technological breakthroughs as significant as any which astounded and transformed the dreams of the generation after 1870. In that single generation, virtually everything in the above paragraphs was either invented or outlined in principle—the last being the television set in the early 1920s.

Peter Drucker, in his book *The Age of Discontinuities*, discusses the kind of phenomena which occurred from 1870–1920, arguing that life goes on in a relatively understandable and predictable continuity until, suddenly, *everything* changes. Such an all-pervasive change is caused by so many things being invented and changing that the collective impact can only be understood as a "break point." On one side of this break point is the world we have always known and understood; on the other side of the break point—a consequence of so much invention and discovery—is a new world which has its own framework, assumptions, and habits.

In *The Age of Discontinuities*, written in 1967, Drucker claimed we were entering a new era in which the ideas and assumptions of the post-World War II era would fall by the wayside—and simultaneously asserted that one could not predict in detail what would be on the far side of that discontinuity: by definition, if its results were predictable, it would not be a discontinuity. However, Drucker accurately predicted the decay of the steel and automobile industries as well as the rise of the service and information industries, and he described the pain of workers accustomed to one way of life who find themselves in dying jobs and dying towns while their children move on to new towns and industries.

If you spend an evening with Drucker's 1967 book and with John Naisbett's *Megatrends,* written in 1982, you notice that both books give varying descriptions of the same process. Drucker stands at the end of the industrial era and warns that we are entering a new period; Naisbett stands in the middle of the change and plots its pattern. Yet both men are describing the same phenomena, telling of an America in transition from one type of economy and lifestyle to another.

My optimistic vision of the world in which our children may live is very different from our daily reality, but it is actually closer to our world than a description of 1925 would have been to the world of 1870. The coming changes are not as acute as those of that previous discontinuity.

All the possibilities outlined below will be feasible in our grandchildren's world, although some may occur late in their lifetimes and some may not prove worth the effort or be precluded by more spectacular advances. However, the picture of an optimistic future put forth in this book is technologically possible within the next two generations—*if* we have the courage to turn our dreams into reality.

And what are these changes?

Distance will evaporate as a limiting social factor when communication replaces transportation as the primary mode of human interaction. We already see geographic neighbors being replaced by electronic neighbors: your best friend is probably not the person next door but a person you talk to on the phone.

My oldest daughter took a year off from college to live with a French family near Paris and learn about another culture. When I first tried to call her, I was shocked to discover I could dial direct internationally by first dialing the 001 code for the international telephone computer, then the code for France, then the area code for the region near Paris, and finally the code for my daughter's telephone.

Since I had lived in France during the late 1950s while my dad was stationed in Orleans with the U.S. Army, I remember how bad the French telephone system was just 25 years ago: it was so difficult and expensive to call the United States then that only in an emergency did anyone call home. Now, if I get up early enough in the morning, I can expect not only a good connection, but a reasonable rate as well. In less than one generation, there has been an amazing decrease in cost and increase in efficiency in the system.

Our grandchildren, as we view them through our window of opportunity, will be using a complex information-net telephone as easily as we dial the information operator or call our best friends. They will utilize the library by telephone, shop by telephone, send information to and from their workplaces by telephone. As satellites and computers keep bringing down long-distance rates, people will make calls to any spot on the globe—and perhaps to friends on space habitats or on the Moon—as routinely as we now call friends across town.

Many of our grandchildren will do much of their work from their homes by connecting keyboards to their telephones to write letters, books, and purchase orders. Hotels will routinely have full information-set connections in every room. Wherever you are in the world, you will be able to work, shop, and learn.

Note that this breakthrough, as it liberates our grandchildren, simultaneously teaches self-discipline: although they can work while on vacation sunning in the Caribbean or skiing in Finland, they must work effectively without direct supervision.

This decentralized work system will almost certainly mean a decline in 9-to-5 jobs and the re-emergence of piecework. People who love freedom will rapidly adapt to working on their own schedules and being paid for the product on delivery, creating an extremely fluid marketplace of skills and services.

Self-determination—our power over our lives and resources—will also expand due to the explosion in computer information-handling. We already live in a world where you can charge a cup of coffee in the Singapore Hilton in local currency and expect the charge to show up, a month later, on the right bill sent to the right address in the right country and translated into the right local currency.

The coming era of expanded machine-accounting may allow us to build a more free and open society in terms of work, pensions, and government. We might build personal pension accounts which will keep track of us all our lives: credits toward retirement could be transferred when we move to a new job as easily as computers now convert currencies when we buy goods in one country and pay for them in another.

The growth of computer power and microelectronics will lead to a burst of hope and opportunity for Americans who are handicapped or disabled.

My eyes were opened to the great hope the future holds out by a Control Data employee who, after surviving a severe stroke, was left flat on his back facing a future of total disability until his employers loaned him a PLATO system—a then very advanced touch-screen computer. The stroke victim found that at first he could only touch the screen for about three seconds before collapsing in exhaustion. But because the computer is infinitely patient, he could work whenever he liked—at three in the morning or one in the afternoon—and at the pace his body dictated. This man, now completely recovered, last year paid more in Federal income tax than the total amount he received in disability payments in the two years he was incapacitated. And he relates his story with tears in his eyes and a passionate hope for others to have the same chance.

There is an enormous amount of work going on in the area of helping Americans with disabilities or handicaps: experiments in restoring sight to the blind by transmitting light-patterns via computer to the ocular nerve endings, experiments with systems which allow paralyzed citizens to walk by having their minds tell computers how to move the metal supports which hold them up, efforts to develop systems to solve virtually every problem that limits human mobility and sensory perception.

Today, by contrast, disability is handled as a welfare program that offers a very small maintenance income for people who can no longer do their former jobs. Because we define it as a passive support program, there is cheating and malingering which has resulted in government efforts to screen the program tightly and to double-check everyone. This leads to a frightening life for people on disability, some of whom are not very well educated and have marginal but very real disabilities, as they are pushed and prodded by a government they thought was there to help and protect them.

It may be that our grandchildren will not fear life's ills and disabilities but be able to accept them as an inevitable component of longevity, assess their capabilities at each stage of life, and retrain themselves as necessary. The technology is here today with which we can begin this badly needed development program.

In the optimistic future we are exploring, our grandchildren may link biology, computers, and micro-electronics to help those Americans born with retardation and birth defects. We have no notion at present what benefits might accrue by interfacing a mentally-retarded child and the right computer system. We do know all too well the limitations of our current support and teaching systems in dealing with people who are retarded, but there is a world of discovery awaiting us—*if* we view the plight of these disadvantaged children as an opportunity for research and development beneficial to us all.

Many of the opportunities we glimpse through our window have little to do with developments here on Earth. One of the great revolutions in our lifetime has been man's leap beyond the planet. As Carl Sagan is fond of noting, for all the rest of human history there will have been only one magic moment when we first left the planet, braved the near-vacuum of space, and reached our nearest neighbor, the Moon. That moment, of course, belongs to our generation, and the people who did it were Americans.

We Americans have had a tendency to be indifferent to space, to look at the space shuttle and at potential space stations as having abstract scientific and technical merit but no direct impact upon us here on Earth. It is becoming increasingly clear to some of us in Congress, as well as in the private sector, that an American commercial presence in space has the potential to create heretofore undreamed-of opportunities.

Congressman Bob Walker of Pennsylvania has been exploring the possible benefits of weightlessness to people currently restricted to wheelchairs. In speeches to handicapped Americans, he makes the point that in a zero-gravity environment, a paraplegic can float as easily as anyone else. Walker reports that wheelchair-bound adults begin asking questions in an enthusiastic tone when exposed to the possibility of floating free, released from their wheelchairs. Several have volunteered to be the first pioneers. In fact, the space shuttle already takes off and lands smoothly enough that a person in a wheelchair could easily be aboard as a mission specialist.

This is only one example of compassionate high tech. If given the opportunity, I am certain that our grandchildren will combine space flight, work under the oceans, computers, microelectronics, and biotechnology to make the problems of disabling handicaps nuisances rather than life-shattering burdens. That future is no more amazing than the generation of 1870 would have found a world without polio, typhus, yellow fever, malaria, and cholera.

Not only Americans with handicaps—but, indeed, all Americans, may have great revelations of hope and opportunity ahead for them. In space alone there are many possibilities which remain unexplored because of the negative mindset of the welfare-state bureaucrats, the limited budget of NASA, and the hesitancy of our aerospace companies to invest in large-scale independent research and development programs in the absence of either strong government commitment or public support. Since 1969 and the landing on the Moon, we have been waiting for someone to give us a new vision of our purpose and role in space. We are *still* waiting for such a visionary, someone able and willing to translate technological capability into human opportunity.

If visionary thinking succeeds, by our grandchildren's time, we will have factories in space—large structures, easily sustained in the weightlessness of near-vacuum—producing medical goods, special alloys, vacuum-formed surfaces, and a host of other materials.

In medicine alone, we may find that the effect of weightlessness on certain manufacturing processes carried on in the relatively sterile and pure environment of space will result in a multibillion-dollar industry. Johnson and Johnson has joined with McDonnell-Douglas in a series of experiments which indicate that within a decade there could be a very profitable industry making medicines in space to be exported to the entire planet.

The materials industry is very interested in the prospect of alloys composed of lead and aluminum which can be formed only in a weightless environment. It is possible that, in a new partnership with nature, we will presently be discovering metallurgical characteristics impossible to create in Earth's restricting gravity. Furthermore, it may prove feasible to grow very large crystalline structures in space, a process which, by itself, will yield manufacturing breakthroughs literally unimaginable on Earth. It is important to note that we are not talking about simply doing something a little better than we can do it here on Earth. We are describing processes which are actually impossible while Earthbound.

It has been a sign of our lack of imagination that, while many people talk about Japanese and other foreign competition, there has been no concerted effort to build on the unique monopoly we have—for the *present*—on developing factories and jobs in space.

Here is an area in which our lead over the Japanese and the Europeans is so extensive that, if we simply build on it, our grandchildren will have an enormous export advantage. We have used large government programs before to build these kinds of export advantages: throughout the post-World War II era, the United States enjoyed an enormous advantage in commercial airliners and in computers because of the research and development commitments of the U.S. government in general and the Department of Defense in particular.

If we will make an intensive effort to develop space, we will create millions of jobs on Earth while creating thousands of jobs in space, while at the same time ensuring a solid balance of payments in foreign trade by producing goods and services others want but cannot produce for themselves.

On a lighter note, Herman Kahn once suggested that the biggest growth industry of the 21st century will be space tourism. As people grow wealthier and the cost of space transportation comes down, spending a week's vacation on a space station or a honeymoon on the Moon may become commonplace.

Space travel today is where air travel was in the early 1920s. One reason I am convinced space travel will be a growth industry is because I represent the Atlanta airport, which provides 35,000 aviation-related jobs in the Atlanta area. Delta Airlines, which has its headquarters there, employs some 15,000 people and is the second largest private employer in Georgia; Eastern Airlines has its largest operations center there also.

Both companies started with small airplanes: Delta started with two crop dusters and, through an enormous effort led by C.E. Woolman, survived the Depression to become a great company; Eastern Airlines began as a very small company using tiny airplanes and, because of the genius of Captain Eddie Rickenbacker, grew into a transportation giant employing thousands.

Consider that the entire first flight of the Wright brothers' airplane—take-off and landing—was no longer than the wingspan of a Boeing 747 and you can begin to appreciate how far aviation has come in eighty years.

Early airplanes were far cruder and far smaller than the current-generation space shuttle. They were unreliable, unsafe, and uneconomic. Private airlines could not have survived at all without government subsidies underwriting the building of airfields, the Civil Aeronautics Board, and government-bonus rates for airmail.

Because American politicians had the vision in the 1920s and 1930s to subsidize and nurture an infant industry, we came to lead the world in aviation—a leadership which defended not only the U.S. but also our Allies, and which saved countless American lives in World War II.

Today's space shuttle is only two generations short of being the DC-3 of space travel. We will have to build one more transitional space shuttle system, probably in the mid-1990s. Although this system will carry more passengers and more freight, and also be substantially cheaper than the current shuttle, it will still be too expensive per passenger or per pound to make space travel economical for the average person. However, a third-generation space shuttle—built with the new materials, computers, and propulsion systems of the 21st century—*will* be the DC-3 of space. From that point on, people will flow out to the Hiltons and Marriotts of the solar system, and mankind will have permanently broken free of the planet.

While space is an important area of growth for the next generation, it may not be the most dramatic: the greatest changes ahead may come in the area of biology.

Biology is entering a period of intellectual flowering like that which transformed physics after the turn of the century due to the seminal contributions of Einstein, Bohr, Heisenberg and others. The discovery of the structure of DNA by Watson and Crick is symbolic of a new era in biology which may reveal the very secrets of life. While most of our attention is focused on microbiology and efforts to develop synthetics and new hybrids at the micro-level, developments are also under way in ecosystem modeling, ethology, and a host of other areas which will allow us a deeper understanding of the organic world by early in the next century.

This biological revolution may enable us to feed a world with a population vastly greater than that of any current projection. It is a comment on the scale of our agricultural capability that virtually every developed country already produces too much food.

Even Japan—the most heavily populated-per-square mile of the developed countries—protects its farmers against foreign competition.

The United States, Canada, Argentina, Australia, and New Zealand have classically been the world's agricultural exporting countries. They have recently been joined by the European Common Market, whose agricultural system is so efficient that it is building mountains of milk, butter, and cheese and shows signs of swamping the world in chickens and eggs.

The Third World's problems with food production are actually problems of education, capital investment, and government policies which keep farm prices depressed so that its urban poor don't riot. If Third World countries would adopt the policies needed to encourage farming, we would have a massive food surplus worldwide and we would recognize that hunger is a political-social problem and not a resource problem.

The next two generations may see such significant breakthroughs in our understanding of the biological world that we will develop a pro-environmental organic model of farming which will improve and enhance nature while producing nutrition for humans. The University of Georgia's work on various non-till systems of farming is already having an impact on soil erosion. Others are developing biological rather than chemical methods of pest control and better types and applications of fertilizers. Serious research is under way to produce new strains of wheat which will function as legumes, actually putting nitrogen back into the soil. These methods will have far less negative impact on the environment than did more primitive systems of plowing and fertilizing.

The biological revolution will transform far more than food production. Already there have been enormous advances in health care, and the lessons we are learning in microbiology may soon remove cancer from the list of life-threatening diseases.

If we are going to retain our high standard of living and compete in the world marketplace, Americans must give priority to the development of high-value industries such as space tourism and advanced health care. Today and henceforth, anything which is simple can be done at less cost somewhere else on the planet. However, those systems which require long-term investment in human education and sophisticated equipment will hold their value even as the less-developed nations become our competitors in low-technology fields. Thus we should be emphasizing for our grandchildren jobs and opportunities in new high-technology industries.

One of the lessons of biology which will almost certainly color our grandchildren's lives (as it has already begun to shape our own) will be the dicta that ultimately our bodies are organic and must be respected.

The exercise boom is the first major step toward the form preventive medicine may take in years to come. Our grandchildren may invest time and money in avoiding illness by staying in shape; they will recognize that man's greatest enemies are no longer the viruses of the past, but our own behavior and habits.

There is a growing market of future senior citizens who believe that their own habits can determine whether they spend their last years in a nursing home or in productive pursuits. As we discover that understanding and manipulating the biology of aging will allow us to remain young far longer, people will begin to develop a different outlook on the prospects for life in their eighties and nineties.

As yet, we haven't been able to extend human life beyond the limits enjoyed by the patriarchs of the Old Testament, but we are rapidly increasing our ability to extend the average lifespan and to avoid premature death.

As lifespans of longer than seventy-five years become commonplace, our children must inevitably change their image of retirement. Most people have no wish to spend twenty-five years or longer in enforced inactivity. The desire to remain active will increase as jobs become more knowledge-oriented and less physical. Formerly, people who worked in factories and farming were simply physically exhausted by the time they reached their sixties; the cultural necessity of retirement came from jobs which literally wore out workers.

In traditional societies—before heavy plowing or industry—people stayed active all their lives; as they matured beyond hunting or detail work, their focus shifted to teaching, advising, and other special skills which an older person acquires, but which requires no great physical effort. There were no wasted years during which people were warehoused while waiting to die.

As jobs evolve away from physical labor, more people will remain active and productive because their work interests them. It is almost impossible to be effective in an information-age job if you don't like your work. One of the keys to mastering any information-age profession is simply the amount of time put in: a professional golfer hitting five hundred balls before breakfast and a lawyer taking his legal reading on vacation to catch up while resting are real professionals getting paid for playing games they enjoy.

It is this combination of profit and pleasure which will lead many information-age workers to want to continue working well past their "prime."

Today our society actually discourages this kind of positive behavior. Nothing bothered me more, as a Congressman, than our government's treatment of Warren Brockway from Newnan, Georgia: Warren, a retiree, wanted to do something to keep his mind and body busy and to contribute to the community and world he loves, so he took up scrimshaw—ivory carving. Once he became so adept at scrimshaw that he could sell his beautiful work and earn the money to buy more ivory, the Social Security Administration told Warren that, if he did not stop selling his art, his Social Security benefits would be cut off. Deciding he could not afford the economic penalty, Warren was forced to give up selling his art. But Warren's need to serve could not be stilled—today, at 71, he is head of the Newnan Public Housing Authority.

Our children and grandchildren will radically transform our current retirement rules as they rebel against the anti-activity bias of our current system and demand that we encourage, rather than discourage, activities after sixty-five. This they will do in part for their own mental and physical well being, and in part because they will recognize that staying active is the key to staying mentally and physically healthy.

However, there is a second reason for our grandchildren to stay active throughout their lives: the harsh reality is that no society can afford to have so many people as will soon be seventy-five or older spending more than the last third of their lives dependent on a smaller working population.

This generation of adults will be the generation which forces the issue of social security reform and mass retirement reform. Despite all the demagoguery of Democrats over the last few years, the fact is that the system they built forty years ago cannot endure.

There will be too many baby-boom (people born between 1946 and 1964) retirees and not enough children of baby-boomers to let all of us retire for the last third of our lives. Many of us will not have saved enough to supplement social security and our pensions and make us self-sufficient. Our children and grandchildren may simply refuse to give up their take-home pay and their dreams of homes and vacations in order to subsidize the enormous number of older Americans who will eventually represent the largest population bulge in American history.

Our children and grandchildren may well have a more aggressive and activist view of both ends of the age continuum: just as they will be more active in their later years, they will participate more in their youth.

I first came to realize how much the information age will change our vision of youth when I was teaching a Federal executive seminar in Washington last year. One of the seminar's members owned a small software company in partnership with her husband. The couple got summer-job applications from twelve- and thirteen-year-olds, young people with home computers who were adept enough in their use to be employable at a much higher wage than the average teenager could expect to earn from a menial summer job.

Recently I heard a story about an eighth-grader in Florida who built a successful business as a weather consultant serving large corporations by processing free information—information which the government had on its computers but which normally could not be processed for two or three months. This young man, who was doing the processing and sending out weekly reports via electronic mail to over forty national corporations each week, was making money by providing his clients with a valuable service for less than they could obtain it elsewhere.

It is conceivable that, by our grandchildren's time, adolescence will have become a thing of the past: future historians may conclude that adolescence was an invention of 19th-century parents designed to keep their children out of textile and steel mills.

If young people enter the labor force earlier, there might be a revival of the concept of apprenticeship. Youngsters could be paid a modest wage to work and study under a master of their chosen trade. Anyone familiar with Benjamin Franklin's autobiography or the pre-industrial lifestyle will recognize that there is much to be said for learning while working.

Once we accept the notion that an information society may require us to work all our lives, we will change the whole structure of education. After all, once you have learned the basics in a controlled environment, it may be more effective to let you seek the knowledge of your choice at your own pace rather than trying to organize knowledge and present it to you in homogenous form at a median pace.

Our grandchildren may be able to look back in wonder at the degree of specialization—a natural product of the industrial society—which caused our lives to be organized into learning years, working years, and retirement years. Future generations may well enjoy learning, working, and vacationing every year—

provided they live in a society in which people negotiate for the results of a job instead of the process of the job. The new flexibility of this information-age work ethic, combined with the physical freedom brought about by the telecommunications revolution, can bequeath to our grandchildren greater freedom in ordering the very substance of their lives—their time.

One of the greatest breakthroughs in our children's world might be this very different approach to life, structure, and habits. To make this new world of our children a reality will require self-discipline on the part of each citizen. Decentralization will, of necessity, limit our ability to enforce behavioral codes upon others. Technology will permit individuals to work and learn at home, but there will be no supervisor or teacher present to enforce productivity. The worker or student will find his attention focused and his time disciplined by the goal he has agreed to achieve.

In this bright future, our children will be comfortable with change and the need to adapt to their fluid society; they will accept as a given that whatever job they do at twenty may not exist when they are fifty. They will invite and enjoy constant change.

There was a period after World War II when our society had become so successful it presumed to dictate terms to the world around us: we decided that a man could choose one career and that forever after the world owed him a living, even should that career become obsolete; we decided that population centers had a right to exist even if their function in the marketplace disappeared. We have wasted enormous amounts of time and money propping up these outmoded assumptions and trying to avoid the inevitable changes which occur in all human society in the real world, where we have remarkably little resistance to the constant pattern of invention, innovation, and adaptation which has been the signature of Western man for four hundred years.

Imagine how different America's 19th-century industrial revolution would have been if we'd had a welfare state then: the stagecoach drivers' union would have protested the development of railroads and argued vigorously for subsidies for the past (their industry) and against the future (railroads). If this sounds far-fetched, consider the number of special interests that have organized politically to avoid the necessity of change, that are taxing the rest of us for the privilege of remaining who and what they've always been, without regard to the hardships this causes others.

Our grandchildren will be comfortable in a world where every business, community, and citizen must adapt or pay the price of obsolescence. They will value successful change and successful adaptation because there is a deep need in the younger generation to affirm their intuition that there must be more to life than standing in line at the steelworkers' union hall for unemployment checks or getting drunk at a local bar discussing the unlikely chance that their steel mill may reopen.

Rather than whining that change is frightening (which it is), that change often dissociates us from our roots (which it does), and that change sometimes has undesirable side effects (which is true), our grandchildren will accept these caveats as facts of life. They will ask their culture to teach them how to rise above these difficulties rather than to hide from them.

This comprehension of the inevitability of change will bring with it an impatient demand for an efficiently decentralized society. Today, all large organizations—whether corporations, labor unions, professional associations, or government bureaucracies—are hampered by their unwillingness to change, and yet they survive, forestalling the inevitable, by using their very size as a weapon with which to hold the forces of adaptation at bay.

When an institution is large enough, it can shape its own reality for a time: big companies outlast their competitors by using their influence to rig the rules of the game, thus securing special advantages; trade and professional associations enact laws to keep competition from forcing change upon them; in extreme cases, huge systems can extort direct subsidies to prop them up, even though the goods and services they produce no longer serve any real need.

The United States has actually been relatively lucky in its decentralization when compared to Western Europe. European farmers and industrial workers are far more tightly organized and more militant, willing even to blackmail the rest of their society into paying them to do nothing.

The result in Europe has been disastrous Common Market policies in farming and in the iron and steel industries. The more Europe's politicians provide subsidies to avoid change, the more its business and labor leaders learn to focus their efforts on forcing the political system to underwrite the status quo.

If Western European labor and business leaders concentrate on extorting money from their government rather than finding ways to adjust in order to compete with the Japanese and Americans in our rapidly changing global marketplace, they will fall even farther behind. Furthermore, since by now they have been taught that their governments should take care of them, they will blame any bad results on those governments. Thus, we have a vicious circle: by focusing labor and business on the wrong arena—politics instead of marketplace—an ever-worsening pattern is established.

Great Britain is the country which has most clearly followed this path of decline and decay by leaning on its increasingly socialized government instead of letting the marketplace force adaptation and change. As successive British administrations propped up industry, it became less able to compete with the Americans, Germans, and Japanese. Britain, the most industrialized nation in the world in 1880, is on a par with Spain in annual per capita income for 1983. Only with Margaret Thatcher has that decay been directly challenged.

Our children and grandchildren, having seen the costs of subsidizing those who will not change, may take the position that society owes its members only the *opportunity* to change, *not* the right to avoid change.

The retraining and reinvestment programs of the next century will be gigantic by our standards. An information society requires as much investment in adult retraining and re-education as the industrial society required in public schooling. The greatest single problem we face in trying to manage the coming transition is that of making it relatively easy for people to adapt.

As long as we support a laissez-faire policy toward adaptation and change, people will resist bitterly being propelled toward what will seem to be a strange, unknowable future in which they may have no value and no part. Imagine the shock of learning, at the age of forty-five, that, although your grandchildren may do well, you are unemployable and irrelevant. The welfare-state policies of subsidizing structural unemployment and keeping people in decaying industries by giving them just enough to subsist has been bad enough, but pushing those people into a brutal life of change-or-die would have been worse.

No society as wealthy as ours should allow people to suffer simply because they took the wrong job; or the right job one year became the wrong job a decade later; once a system for continuing self-education is developed, our able-bodied should be required to throw themselves into the game of life, instead of sitting on the sidelines with subsidies in their hands.

If the bright future pictured through our window of opportunity is to become a reality, our grandchildren must develop a continuing self-education system which combines the image of compassion for which Franklin D. Roosevelt is known with the emphasis on productivity that we associate with his older cousin, Teddy. In many ways, Teddy Roosevelt might prefer our grandchildren's America to ours—an America comfortable with the assumption that people must try new things all their lives because any particular occupation may become obsolete in a single decade.

Thus, while requiring a strenuous commitment from

every citizen which Teddy Roosevelt would have approved, our grandchildren may create structures and incentives for organized learning and adaptation that F.D.R. would have applauded.

Because of their necessary commitment to change and adaptation, our grandchildren may well have high expectations of the public and private systems which serve them, little patience with arrogant businesses or aloof government agencies which seek to avoid change and then ask customers and citizens to pay the cost, and a strong bias in favor of small management systems and big systems that behave like smaller systems.

For example, our generation believed that the answer to big business was big labor and big government. Experience has shown us that all three of the "bigs" tend to ignore the average person and pursue selfish interests. Even worse, the three big systems collude to share power at the expense of the general public.

Thus, the populist-progressive hope that big government could master big business and the Liberal hope that big labor could counterbalance big business both proved false. As a consequence, recent Liberal efforts to protect the public from big business have met with indifference: relatively few people really believe a large government bureaucracy devoted to consumer protection would actually serve them; instead, the public suspects that any such bureaucracy would rapidly be co-opted to serve the very interests it should police.

One step toward returning power to the people in a realistic manner might be a small claims court that could allow consumers to litigate suits under $10,000 without the expense and inconvenience of retaining a lawyer. Such a night and weekend court, based on mediation and arbitration, should be designed for laymen rather than as another way to enrich the legal profession while delaying and distorting justice. People need to know that, if a large corporation rips them off, they can pursue justice without incurring huge legal fees and losing days of work. Such a court needs a practical business sense and a calendar which allows every case to be heard within a month of filing.

By having the loser pay court costs, the system could be self-financing and consumers with nuisance claims kept to a minimum by the knowledge that losers must pay court costs; at the same time, those with genuine grievances would be assured that, even if they lost, they would pay no more in court costs than an attorney's fee would have cost in the older system.

Such a consumer court would provide computerized instruction designed to teach citizens how to prepare and plead cases without complicated legalese. Based on the notion that justice should be available to all, it would reject the idea that laws and procedures should be so arcane, even in minor cases, that only a full-time attorney can master them.

By teaching every citizen the basics of courtroom procedure and of justice, these consumer courts could re-establish the bond between citizen and law. By being available to any citizen, such courts would greatly enhance the power of the individual.

Another innovation that could increase the power of each citizen is the creation of a "government services personal credit card," which could change the entire balance of power between citizen and government bureaucrat.

At present, the Federal government sends large chunks of intergovernmental funds to state and local governments, money which goes from Washington to the state capital or local government in a bureaucrat-to-bureaucrat transfer of taxpayers' money.

The creation of such cards would return customer status to the citizen and create a free market in local governments: the Federal government, in a form of revenue sharing, would divide its intergovernmental monies into allotments due each citizen. In January, every American would receive the year's credit card and credit line, which they would be empowered to use, either in person or via information-set telephones, to indicate the goods and services they wished in the coming year; the government, in turn, would fix prices on these services and charge them against the cardholder's credit balance.

Consider the effect of this innovation on a public housing project of five hundred citizens, each of whom had a government personal-services account containing $600. Their housing association would have the potential to commit some $300,000 in funds, allowing them a greater say in local government than has ever before been possible.

At the same time, such a power base would encourage cooperation between citizens groups and local government and responsible decision-making at every level in a social structure more participatory than it has been since our nation was comprised of thirteen colonies, as well as heightened levels of self-discipline and citizenship of the sort necessary to our optimistic future.

Our grandchildren will be more demanding not only of local government, but also of themselves: the emerging information-society environment will offer greater flexibility and freedom, but it will also demand foresight and self-discipline.

The industrial age developed a contract between owner and worker in which the owner took on responsibility for management and the worker did the labor: the owner took care of everything except the immediate job and the worker showed up for forty hours a week and did as he or she was told. This style of management reached its peak in such classic heavy industries as steel and automobiles.

The last twenty years have eroded this notion because Japanese-style teamwork is proving to be a more powerful production model. At the Hapeville Ford plant in my district, I have participated in joint management-labor efforts to create quality circles utilizing a management-labor "team" approach that would have been anathema to *both* Henry Ford and Walther Reuther.

After several years of layoffs, automobile workers in my district have come to believe that their commitment to quality and willingness to cooperate with management may be the key to keeping their factory open. We are moving toward a wider distribution of management responsibilities in which every employee shares the burden of seeing that the company survives.

There has been a similar shift in attitudes at Eastern Airlines. Over the last fifteen years, I have talked with hundreds of Eastern employees about their company and their attitude toward management. Under Captain Eddie Rickenbacker, Eastern had an effective—if authoritarian—management style, but the management after Rickenbacker was unable to direct the company effectively and Eastern became a troubled business which incurred huge debts. This burden of debt made it difficult for Eastern to compete with relatively debt-free Delta Airlines. Furthermore, Delta's management style—that of an open door at every level and a deep commitment to teamwork—gave the Delta family (as they call themselves) real psychological advantages.

Now Eastern is taking a step toward real teamwork: in order to finance its debts, the company has asked its workers to give back a substantial amount of their salaries, simultaneously agreeing to sell them stock in the company in return for their commitment.

The result is a major airline, 25 percent of it employee-owned—a major step toward self-management. From now on, decisions about the company's future will involve workers both as employees and as stockholders.

The Eastern experiment is a step toward the joint worker-owner style which will dominate the information age. When the quality of work is critical, it is essential that workers have a proprietary rather than an adversarial attitude toward their company. Gimmicks and public relations efforts will not achieve this sort of commitment. People bright enough to be effective in the information age are going to be too bright to be conned by management tricks.

Because our grandchildren will live more in the Sunbelt than they do now, their lifestyles will become increasingly outdoor oriented: physical fitness will continue to be a major growth industry because, as people grow wealthier and more responsible for their own well-being, they will have both the opportunity and the inclination to stay in good physical shape.

It is likely that our grandchildren will prefer more leisure to more money and will use that leisure to improve their bodies as much as their minds.

The physical labor of the past is being inexorably replaced with mental work or with service occupations such as word processing; because of this, we will increasingly desire to use our bodies to release the day's tension by lifting weights, swimming, or through contact sports.

Many of our grandchildren will choose careers that blend lifestyle with workstyle, mental exercise with physical exercise—the antithesis of the industrial society, in which forty hours of labor were surrounded by free time spent in ways completely unrelated to the assembly line or the blast furance.

This new approach to living is being expanded by technologies which permit people to work at home. Home occupations have always existed, to some extent, in America, but the new possibilities inherent in advanced technology will soon allow millions to work at home. Already, Lanier and other computer firms are experimenting with work stations in the home by using telephone lines to carry information to the worker instead of highways to carry the worker to the information.

Initial experiments indicate that this style of work will require social innovations to make it successful: we must learn to pay people fairly for their productivity on a piecework basis and allow them to create their own working hours.

People will contract to produce 10,000 words of material or per day or to do a particular project over a week, month, or year, and the employer will not concern himself with how or when the work is done—while the baby is asleep or late at night—because the employer is buying the product, not the process.

Working people will find no single change more valuable in increasing their leisure time and real income than the development of work stations in the home. Consider, first, the issue of hidden costs. Today, most working people pay for transportation to and from work, parking while they work, and a commercially prepared lunch. Deduct these hidden costs and take-home pay decreases drastically. If we further deduct the cost of dressing for work there is an additional decline in real income.

Today, big companies deduct the costs of providing a work environment; tomorrow, when people work at home, they will be able to benefit personally from deduction for work rooms and work-related expenses that increase their after-tax income.

Our work force will, however, gain benefits in lifestyle which outshine those in income when work stations are common in American homes. Today, millions spend hours in their cars or on mass transit commuting to and from work. Because work is currently a process rather than a product, and because most people work the same hours, we endure seemingly inevitable traffic jams and surges of peak traffic unrelieved by our attempts to create freeways and augment mass transit to meet our growing needs.

People who work at home gain an average of an hour of travel time, each way, in a metropolitan area that amounts to the equivalent of twelve weeks per year. These extra hours of new free time gained by working from home can be devoted to earning more money or used for leisure.

Neighborhoods will be revitalized when more people are in their homes during the day; there will be fewer burglaries, less crime, and a significant increase in neighborhood businesses as well as a return to utilizing the neighborhood as a community center.

More importantly, the family unit will be strengthened by the shift to working at home: once you control your own time, you are free to schedule that birthday party or Cub Scout meeting which used to be so troublesome.

Male and female roles will be redefined and old values rediscovered: on the family farm in the agricultural era, the family worked together, ate together, and prayed together. Then, in the era of heavy industry, the male or both adults went off to work and focused intensely on demanding, sometimes brutal, labor in order to earn a living.

Home work stations will allow both partners to share in parenting and family chores and permit a far more flexible lifestyle in which self-directed work choices rather than the coercion of an inflexible employer-dictated routine dominate our lives.

Despite their greater range of choices, I suspect that our grandchildren may be more conservative and traditional than we have been. The religious revival which is under way across this country has an intensity which is beginning to resemble that of the Great Awakening of the 18th century.

Societies seem to go through cycles in which one generation rejects the preceding generation's experience. Whether out of boredom, children's natural reaction to their parents, or a new set of lessons learned from the mistakes of our elders, an almost organic flux seems to be at work: societies develop waves of behavior; entire generations seek new habits, approaches, and definitions of life.

Today it seems that we are moving toward a new-time religion, rather than an old-time revival, very much as did those citizens in the era of the Great Awakening.

Let's look closely at two instructive periods in history which help put this transition in perspective: the 18th-century Great Awakening in both its American and British versions, and the dramatic transformation of England in less than a decade from a hedonistic Regency to the egalitarian conservatism we associate with Queen Victoria.

There have been few periods in history that changed a society as dramatically as did the Great Awakening in the English-speaking world during the first half of the 18th century. In America, there were mass meetings throughout the colonies as revivalists, led by George Whitfield, aroused and instructed attentive audiences. In Britain, the less eloquent but better organized John Wesley created a movement whose methodical approach led people to call it Methodism.

A number of historians argue that it was the passionate redemptive call of Methodism which saved Britain from the revolutionary fervor which overtook France: where the French populace sought equality in a political utopia, the British sought salvation through God as revealed by the passionate missionaries of the Methodist movement.

It is clear that the citizenry which was drowning itself in gin in 1700 was, by 1800, more sober, stable, and productive; it is also clear that people found in religious life an ordering principle which allowed them to bear the disappointments and dislocations of the industrial revolution patiently.

The Great Awakening in America was just as profound, though less organized than in England. Thousands of Americans attended camp meetings and revivals at which they learned that God loved them, that they had real meaning in the world, that they had a moral obligation to stand up for what is right, and that giving in to evil was a fate far worse than death.

The endurance which Americans showed at Valley Forge, as well as their indomitable will in the face of repeated defeats, resulted from the spiritual strength which grew out of the Great Awakening. Britain lost the Revolutionary War because it was fighting a moral, not merely a military, force when it engaged Washington's army.

Despite attempts by today's anti-deists to abolish God and religion in public life, it is impossible to study the leaders of the American Revolution without being profoundly impressed by their commitment to God and to divining His will. Even our most radical founding father, Thomas Jefferson, wrote in the Declaration of Independence that all men are "endowed by their Creator with certain unalienable rights." Thus the very first document by which we proclaimed our freedom from British hegemony asserted that this freedom was granted to us by God and not man.

Any serious look at the use of prayer to open and close meetings, the language on our Great Seal (*Annuit Coeptis*: He [God] has favored our undertakings), and the personal correspondence of many of the founding fathers shows that they believed deeply in the message of the Great Awakening, that they found strength in the knowledge that they were fighting for divinely mandated freedoms and building a system based on rules which came from God and God's nature, and that the natural order in which they all believed was transcended by a Creator.

If the 18th-century revivals in America and Britain had a profound impact on the cultural and religious processes of the time, the same was true of the sudden cultural transformation which took place in the late 1830s in Britain.

After the Napoleonic Wars, British aristocracy developed a social style which was increasingly daring and hedonistic. One young man who followed the style and sudden transformation of his culture was Benjamin Disraeli. As a rising young poet-novelist, Disraeli led a fast life, wore the outlandish clothes of his well-to-do contemporaries, and engaged in a variety of fashionable affairs such as touring Europe with his mistress and her husband and living with them in a *ménage à trois* considered acceptable, even commendable, by the young aristocracy.

Suddenly, things changed and Disraeli's behavior, once a positive force augmenting his social ambitions, became a liability. With the accession of King William IV in 1830, Disraeli's society was transformed: for a variety of reasons, the British citizenry allied itself with the socially conservative new king to create a mood of morality and stability, perhaps because—as some have suggested—the rising industrial and merchant classes wanted a more stable world for their children and a lifestyle compatible with the work ethic which had made them successful.

With the accession of Queen Victoria in 1837 at the age of eighteen, the die of conservatism was cast:

her very youth made it inappropriate to act vulgarly or with the display of outrageous style which had been acceptable only a few years previously.

Disraeli, caught unprepared by this sudden conservative shift, found his political career crippled for several years by his reputation. Had he continued to live publicly with his mistress and her husband, he would almost certainly have been unable to recover from the excesses of his past. But, unlike many of his foppish contemporaries who lacked Disraeli's brilliance and were ruined politically as a consequence, he was able to make a successful transition, adjust to the new order, and become a respected voice in the stable and conservative Victorian era.

Those Americans old enough to remember the stability of Eisenhower's last years in office and the sudden emergence of the Free Speech Movement's radical public-demonstration style just four years later at Berkeley can appreciate how rapidly societal styles change. No politician in 1960 could have imagined the riots in our large cities, the mass demonstrations on campus, or the violence in Chicago during the 1968 Democratic National Convention: in eight short years, the behavior of an entire generation changed radically.

Our grandchildren may well live in a society whose culture and mores also will have changed greatly. One driving force in the coming cultural reaffirmation of traditional values associated with our optimistic future will surely be the "graying" of the baby-boom generation. This generation, born between 1946 and 1964, has been demographically represented as the "pig in the python"—a bulge in the population, a group which, by its very size, stands apart from the population as a whole.

Now that the baby-boomers are entering middle age, it is possible to forecast a shift from their childhood rebellion to the concerns of adulthood: stability, economics, and mortality.

Middle age is the time of life when you begin to realize you are not going to get out of this existence alive. It is a time when you come to grips with the deaths of your parents and grandparents and begin to realize that someday you will be old, even dying. It is a time for stock-taking and for new beginnings.

Thus, as the baby-boomers grow grayer, their numbers may well intensify the religious revival already under way. As the failure of both radical hedonism and economic materialism to satisfy our yearnings becomes increasingly apparent, this intensifying revival of religion could become a new Great Awakening as the baby-boomers swell the congregations of our places of worship.

The rise of high-tech preachers on cable television is accelerating the re-emergence of religion as a legitimate vehicle for explaining the world. Presently there will be religious software for home computers and a host of modern high-tech efforts to spread a new, electronic gospel, an outreach building not only in Christianity today, but in Orthodox Judaism and Islam as well.

Polling data indicates that the younger Americans following behind the baby-boomers—those born in the late '60s—comprise the most conservative generation we have seen in a long time, apparently because they have come to the conclusion that there must be a better pattern for life than the self-centered hedonism of the "me generation" which preceded them.

By no means am I suggesting that the next generation will be less sinful or less foolish than the generations of the recent past. The next generation will be equally prone to human foibles; the difference may be that the coming generation will regard its sins as sinful, instead of arrogantly justifying them on egocentric grounds.

One difference between the left-wing secular arrogance of the last generation's intellectual community and what may be the style of our grandchildren lies in the concept and treatment of heroes. When I was growing up, the intellectual community enjoyed debunking heroes: George Washington was no longer a hero because he probably told lies and wanted to use the government for his own enrichment through public works projects on the Potomac; Thomas Jefferson was no longer a hero because he may have had a Black slave as his mistress—clay feet became more important than real achievements.

This example of the welfare-state mindset reminds me of the old adage that "no man is a hero to his valet." But the fact that Winston Churchill drank and relied on his valet to put out his nightclothes did nothing to detract from Churchill's very real heroism in morally leading Britain through a difficult time, as the fact that many men of affairs combine self-interest with national interest is a testimony to their practical patriotism rather than a sign that they are subverting the public good.

The muckraking tradition inaugurated at the turn of the century was helpful in that it reminded us that great men are still men and that concentrations of power tend toward corruption and need to be watched at all times. This same muckraking tradition, however useful as a correction, became a disaster when carried to the extreme exemplified by the American Left's passionate debunking of heroism and traditional values since World War II. That passion is reaching its all-time peak in the efforts of the American Civil Liberties Union (ACLU) to prevent public institutions from recognizing the Christianity in Christmas. The ACLU's attempts to force religion out of our public life have their parallel in the intellectual Left's efforts to destroy the heroes of American history.

I am not suggesting a reversion to the Puritan America of three hundred years ago, but a free and open society must find stability through spiritual as well as secular answers to our current dilemmas.

The social decay and disorder which began with the Free Speech Movement and the violent public demonstrations of the 1960s have been compounded by an overprotectiveness in the legal system which essentially biases the system against victims and in favor of criminals, against the majority and in favor of the minority.

Virtually every poll shows that crime ranks as either the first or second concern of most Americans.

Ordinary people are deeply disturbed by the prospect that their children may become drug addicts, their grandmothers victims of purse snatchers, or their spouses may be mugged at gunpoint.

Since you can walk around many European and Japanese cities without fear, it is obvious that there are ways to make our lives safer.

But we must demand the alterations in our penal code necessary to discourage crime. Every time a judge favors early release of hardened criminals, that judge increases the problem; every time a drug dealer is released on a short sentence or easy parole, the general public should make it known that it will not countenance a legal system soft on crime.

Recently, in response to increasingly vocal public sentiment, criminal sentences have been growing longer in most states and, as a result, the crime rate nationwide has dropped in the last two years.

But with so many people behind bars that our prisons are overcrowded, some states have adopted policies which turn our prisons into revolving-door institutions from which nonviolent offenders can expect release almost as soon as they are sentenced. If unchecked, these policies will soon lead to a revolt by the public.

There is a growing consensus in this country that, if we must choose between locking up the innocent and letting criminals roam the streets, or locking up the criminals and allowing innocent citizens to stroll through their towns and neighborhoods, then the criminals will simply have to be in jail.

This consensus should lead to two new policies in the near future: 1) we must turn prisons into holding tanks for genuine criminals for whom there is no hope of rehabilitation and build more prisons for the permanent incarceration of those dangerous to society—prisons of the "working" variety in which inmates are required to earn their keep because it is simply too expensive to support our criminal population for life; 2) while we must cease trying to rehabilitate hardened criminals, we must simultaneously avoid contaminating nonviolent and first-time offenders by creating more and better half-way houses which will provide correction and punishment for those who have committed minor crimes but can be rehabilitated.

The growing gap between the Establishment Left's analysis of the world and the realities which crowd in on Americans through the new media extends beyond the problem of street crime into the international arena.

The vocal anti-Americanism of the United Nations and of some Third World nations as well as the increase in violent "wars of revolution" are increasingly unacceptable to the average American.

There is apparently a deep-seated fear among Liberals that any critique of the United Nations will permanently undermine its effectiveness. Yet, when Americans watch United Nations debates and United Nations votes which seem to deliberately fly in the face of democratic freedoms and America's commitment to human rights worldwide, we find ourselves increasingly skeptical of the value of an institution which is so "important" and yet so fragile that we dare not discuss its failings openly.

The gap between the average American and the Liberal elite was clearly delineated by their very different reactions to U.S. Deputy Ambassador Charles Lichenstein's recent off-the-cuff comment to the effect that, if the Soviets wanted to pull out of the U.N., we would wave good-bye to them. America's Left was aghast at the notion that we dared call the Soviets' bluff, and further anguished when Ambassador Kirkpatrick suggested on ABC's *Nightline* that the United Nations might want to consider spending half its year in New York and the other half in Moscow.

But average Americans liked both suggestions. Why should we whine and beg the Soviets to let us continue bearing the lion's share of the cost of hosting a patently anti-American U.N. in an open society, when it brings with it shelter for thousands of foreign agents intent on subverting all America stands for? How many Third World ambassadors would reassess their commitment to socialist ideology if they had to contend daily with the KGB (the Soviet Secret police)?

In many ways, the American public has a more realistic view of the United Nations than do many of its spokespersons. As the news media grows increasingly sophisticated in its coverage of the Third World and as Third World dictators grow increasingly arrogant, it is likely that Americans, finding the Liberal apologia unacceptable, will support efforts to reform the United Nations as the only acceptable alternative to withdrawing from it entirely.

The bizarre behavior of Third World dictators such as Idi Amin, Emperor Bokassa I, Quaddaffi, Khomeini, and Baby Doc Duvalier is slowly eroding the anti-colonial good will which was the hallmark of Liberalism's romantic vision of the Third World. From 1945 to 1979 there existed a deep Liberal desire to see only good in the Third World and only bad in right-wing, pro-Western dictatorships and colonial powers. Thus, the terrorism of the Third World's left was always justifiable, while any military or police action by right-wing dictatorships or European colonial powers was reprehensible.

As the news media have grown more discerning in their coverage of the Third World, it has become increasingly obvious that there are events which are impossible to justify or explain why. Idi Amin's antics probably forced the first serious examination of Third World dictators, although the West never really focused on the degree of Libyan and East German tactical support which Amin received. Amin remained acceptable to the Communists despite his behavior (which included cannibalism), because of the strategic importance of Uganda—and to the West, until Ugandans began flocking to Britain and the horror stories told by these exiles began to awaken suspicions that some Third World governments and societies are so accepting of atrocities that we cannot possibly condone their behavior.

The slow development of this view in the U.S. was accelerated by the Iranian hostage crisis, when the full power of American television, newspapers, and magazines was focused on a fundamentalist Islamic revolution intent on insulating the primitive traditions and social sctucture structure of Iranian society from the disruptions of the 20th century.

The Carter Administration was initially stymied in its attempts to deal with Khomeini because it could not grasp the alien nature of Muslim fundamentalism. One of the great challenges to modern Liberalism will be maintaining its sympathetic approach to anti-Western Third World movements in the Middle East despite the knowledge that fundamentalist Islam is anti-female. As Westerners begin to realize that the more devout Islamic movements want not only to keep their women in *purdah*, but also to push them back into the 12th century, the American Left must choose between the Jeffersonian view that all people are endowed with certain inalienable rights and the Liberal determination to view Third Worlders positively, whatever they do, because the Third World is inherently blameless and all its excesses a reaction to colonial oppression.

Younger Americans, who have not grown up with the guilt of colonialism, are likely to look clearly at pernicious dictatorships and interpret their behavior as threatening to Western values. They may conclude that thuggish, brutal tactics are not acceptable even when undertaken by a sovereign government. Something akin to the 19th-century sense of Western conscience is likely to emerge when dealing with particularly barbaric regimes and movements.

We may even see the emergence of new standards of international analysis and behavior. Those Third World nations, such as the Ivory Coast and Gabon, which are moving into the modern industrial-information society in a civilized, non-violent fashion may be regarded as equals of Western nations and treated accordingly; those nations which revert to pre-industrial feudalism and which apply to their citizenry tyrannical controls clearly unacceptable by Western standards may be treated as what they are—either quaint or dangerous anachronisms, depending on their size and their ability to endanger Americans and American interests.

What may emerge during the next decade is a new, American-based internationalism. There are few Americans left who genuinely believe that the United States can withdraw from the world; there are many who believe that our involvement in the world should be based on American interests and American values.

The gap between dynamic, high-tech America and the primitive elements in the Third World is likely to widen during the next decades. Countries like Iran cripple themselves doubly: first, by rejecting Western organizational principles and seeking to re-establish a pre-industrial value system; second, by fighting long and expensive wars with their neighbors that weaken their industrial base and further lower their populations' standard of living.

American exploits in space, computers, biology, and other technical fields may instill in our children a sense of well being sufficient to encourage them to take a hard look at Third World assaults on American values and to conclude both that our system works and is worth keeping and that reactionary dictatorships don't work and aren't worth keeping.

One of Reagan's strengths—from his attempt to retain the Panama Canal to the liberation of Grenada—has been his willingness to embody a reassertive and powerful American nation. He is tapping into a growing feeling that we are a special nation which deserves a firm spokesman unafraid to defend our values and our interests that we have spent two generations too long trying to placate the world. Behind him stands a new generation which reasons that *if* we are the most creative society on the planet, *if* we have the most advanced technology, *if* we continually shoulder the monetary burdens of other nations, *if* we provide the umbrella under which the freedoms of others are protected, and *if* we disproportionately finance the United Nations, then *maybe* other countries ought to listen to us at least as closely as they expect us to listen to them.

This new militancy in America is part of the

same cultural matrix as is our religious reawakening and the restructuring of our legal and bureaucratic systems: the raising of our cultural morale—the willingness to assert our values and stand up for our beliefs and interests—is the first step necessary to realize our optimistic vision of America's future.

This risk-taking, experimental, dynamic future cannot materialize in an America which lacks faith in itself. A strong economic policy and the development of high-technology options require that we become strongly assertive on welfare reform, crime, and foreign policy. People who have achieved success by redefining the work ethic will be just as demanding of their government as they are of themselves. People who have experienced the security born of success in science and free enterprise will want that security extended to protect them, as individuals, from muggers on the street and, as a nation, from the societal muggers among Third World dictators.

Peering into the future through our window of opportunity, I am confident that our children and grandchildren *can* live in a positive, optimistic America—*if* we regain our morale and elan as a society. I am equally convinced that an America confident enough to take risks in space, computers, and biology can regain its capacity to lead the world economically, militarily, and politically.

It is time now for us to turn, in greater detail, to two of the areas which hold the brightest promise for our grandchildren—and for ourselves.

CHAPTER 2: AMERICANS AND THE GREATEST FRONTIER

The most popular series in movie history is the *Star Wars* trilogy. The rush of public support for this romance-adventure-science fiction series changed movie making. In contrast, the film version of *The Right Stuff* has been a disappointment: everyone agrees that the movie is technically well made and carries to the screen Tom Wolfe's version of the original American manned effort in space, but the public has failed to rally to the film.

People clearly respond to a future of adventure, romance, and excitement, but they are not all that interested in docudrama about the technological achievements of man and machine.

The contrast between the ho-hum response to *The Right Stuff* and the smash success of old-fashioned adventure explains a great deal about why America failed to develop the awesome achievements of the Apollo program.

NASA (The National Aeronautics and Space Administration), developed to counter the Soviet Unions's Sputnik during the late 1950s and expanded in response to President John F. Kennedy's challenge to put Americans on the Moon in less than a decade, has been an extraordinary management and technological success.

NASA's achievements are among the high points of all human effort. The Moon landing can be compared only to the Normandy invasion and the Manhattan Project in the speed with which it assembled so many resources, people, and project lines focused to solve so many problems concurrently.

But if NASA's achievement was a managerial

marvel, it was a political disaster. NASA flourished under Kennedy's tutelage and, with the follow-on money his assassination guaranteed augmenting its internal orientation, reached the Moon on schedule, but that same orientation ensured that, once we had reached the Moon, there would be no political constituency to sustain and support an expanded space program.

The NASA team proposed both a space shuttle and a space station to President Nixon who, although distracted by Vietnam and budget fights, nonetheless supported one major program—the space shuttle—as a prudent investment in America's future.

The space shuttle allowed NASA and its major contractors to continue working on new programs and thus was a partial victory. However, the shuttle was a process-oriented, bureaucratic, technocratic program which lacked the romance and adventure that might have created a new generation of engineers and space activists.

Imagine for just a minute a different 1969 in which the President's advisors had convinced him that America had an enormous lead in space and could pull away from every other nation on the planet.

Imagine that the State Department had perceived space-based breakthroughs in communications and energy-generation as the most prestigious foreign aid we could offer Third World countries.

Imagine that the National Security Council had understood that an America which aggressively moved ahead in space would overawe the Russians: the Russians would tread cautiously, ever unsure of what American technical secrets might be brought to bear on them should they indulge in military adventurism.

Imagine that business and industrial leaders had been far-sighted enough to understand that a space industry would spin off earth-based jobs using satellite antennas, new medicines, large surfaces, and zero-gravity alloys.

Finally, imagine a generation of educators who understood that young people need inspiration to motivate them to learn math and science, and that space was the adventure most likely to produce young Americans anxious to master these technical fields so essential to our survival.

With a coherent understanding of the value of space, the President could have scheduled a national address the night man landed on the Moon and, after the world watched that first "small step for man, one giant step for mankind," announced a massive new program to build a permanent lunar colony to exploit the Moon's resources for use in space coupled with the concurrent development of a space shuttle and a series of permanent manned satellites to sustain that colony. Finally, to use the genius of a permanent American free-market system, the President could have announced a series of tax and regulatory incentives to turn space into a profitable arena for economic development.

Given the enormous excitement in 1969 about the Moon landing, the President could have convinced Congress to enact a bold, dramatic space program creating manifold jobs as the program expanded, which, in turn, would have generated a network of labor unions and corporations committed to fighting at least as hard for the space program as dairymen fight for the milk subsidy. There would have been a constant barrage of news covering the development of the lunar base, space shuttle, and our permanent manned station. Young people would have seen real job oportunities twinkling before their eyes in the night sky. Enrollment in math, science, and engineering courses would have skyrocketed.

Where would America be today if we had followed such a bold program?

Our national debt would almost certainly not be significantly bigger. We have wasted the money we might have spent developing space on the welfare state, pork-barrel projects, and mismanagement of Pentagon procurement, all projects without any technical spin-offs in science, job creation, or national prestige.

Had we begun a substantial, ongoing space effort in 1969, our economy would be substantially larger:

one economic study done for NASA by Chase Econometrics in 1975 suggested that even an additional billion a year in space expenditures would have led to 800,000 new American jobs.

Our balance of trade would be more favorable because we would be so far ahead of the rest of the world that the European Ariane expendable launch vehicle would not be competing for commercial satellite launches and the Japanese would not be developing their own launch system. The American lead would have been so overwhelming that other countries would have clamored to cooperate with us rather than compete.

If we had developed at a reasonable pace from 1969, today we would have eight to twelve space shuttles, two manned space stations, and a permanently operating lunar base. Each news magazine would have a section devoted to the week's news from space.

Projects like visiting Halley's Comet could be easily undertaken by a program whose strength and sophistication would be unchallenged; industries—including metallurgy, health care and pharmaceuticals, energy, and communications—would be creating thousands of jobs on Earth for each job in space; America's imagination would have been so sparked by the great adventure that young people would be demanding adequate instruction in math and science to prepare themselves for a dynamic and open-ended future.

But there *was* no bold thrust in 1969. The space program reverted to the bureaucratic shadows and the adventure was delayed while America slept for fourteen years.

Our long delay was not caused by insurmountable technological problems. Nothing we must do in the next twenty years will require a technological expertise unavailable in 1969. If we had launched a crash program with the momentum built by the Apollo program, we would have accomplished more by 1984 than we will have achieved in space by the year 2000. We have been suffering from a failure of political will and imagination rather than a failure of technology.

The power structure in Washington decided to subsidize the jobs of the past rather than invest in the jobs and knowledge of the future. That decision will cripple our nation for another generation if we let it continue, allowing the Europeans, Japanese, and Soviets to surpass us despite our initial lead.

Our politicians are not the first to step back from bold decisions: history is filled with examples of men and women who saw the future and avoided it.

The Ming Dynasty is China is one of the classic "what if" tales in history: in 1400 AD, the Chinese were the most powerful and sophisticated explorers in the world; their ships moved in fleets far larger than anything the Europeans could assemble during that period, exploring as far as Indonesia and Madagascar. It is probable that, had they continued, the Chinese would have beaten the Portuguese around the tip of Africa and met them somewhere on the Atlantic side.

But Ming bureaucrats were very shaken by the aggressive adventurism inherent in true exploration. The process of sailing into the unknown was creating a subculture of adventurous, self-reliant ship captains whose mindset was antithetical to the calm, controlled world of Confucian education and the Imperial Chinese court bureaucracy.

The Chinese regarded their nation as the only worthwhile race and place on Earth. Since they considered all other people barbarians, they had little interest in the world beyond their borders, and thus, to the Chinese bureaucrat, the cost-benefit ratio of exploration was all risk and no profit. Outsiders were barbarians from whom China had nothing to gain, while the act of exploring was creating an unruly class of shipbuilders and owners at considerable risk to the orderly Chinese state.

About the time that Gutenberg's development of movable type (itself borrowed from the Chinese) was revolutionizing Western culture and technology, the

Ming bureaucrats decided to pull their fleets off the seas.

In protecting themselves and their culture from short-term change to which they could have adapted, the Chinese set the stage for the catastrophe which overwhelmed their civilization. The agonizing collapse of Confucian civilization took most of the 19th and the first half of the 20th centuries, killing millions and making the lives of hundreds of millions miserable. The seeds of that disaster were sown by the Ming Dynasty's bureaucratic decision to avoid the future.

History offers brief moments of heroic opportunity. In 1400, the Chinese could have seized such a moment and spread their culture and values throughout the world. But either the will or the vision was lacking, and China has never regained her lost glory.

In the past, Americans have shown a similar willingness to let history's opportunities pass. The decay of the automobile and steel industries shows both management and labor avoiding the obvious requirements for a prosperous future; the U.S. Army's letter telling the Wright brothers that the airplane had no military potential is another example of Americans failing to recognize an opportunity: as a result, the U.S. entered World War I in 1917 with no effective military aircraft despite the three years of intensive aerial combat which had been going on in Europe.

Perhaps the finest American parallel with the opposition to space is Senator Daniel Webster's oration to the United States Senate of the annexation of California on June 27, 1850:

"We all know that it has more than three times as many mountains, inaccessible and rocky hills, and sandy wastes, as are possessed by any State of the Union. But how much is there of useful land? How much that may be made to contribute to the support of man and of society? The agricultural products of the whole surface . . . never will be equal to one half part of those of the State of Illinois; no, nor yet a fourth, or perhaps a tenth part."

Of course, today we know that California is the

premier agricultural state in America. We know that Daniel Webster was just plain wrong.

It was fortunate for America that Ronald Reagan dismissed David Stockman's criticism of a permanently-manned space station: supposedly, in the key final meeting about whether or not to propose a new step into permanent exploration of the potential of space, Budget Director Stockman made the routine bureaucratic objections to spending money in the face of our already massive deficit. Attorney General William French Smith turned to the Budget Director with the comment that the Controller of Ferdinand and Isabella undoubtedly said the same thing when Columbus proposed sailing west to the Orient.

Ultimately, the President proposed a step in the right direction. But it is important to understand just how tentative and tiny a step the space station really is. Consider the testimony of Ivan Bekey, Director of Advanced Programs at NASA's Office of Space Flight:

"The principal difference between the first 25 years of space and the next is that whereas in the past technology limited our grasp, today no new breakthroughs or inventions are required—we can achieve almost any goal in space we set our mind to do. The ability to build and repair in space, provisions of high power and energy, cheap and reliable space transportation, and permanent manned presence in space will allow for fantastic possibilities. Space will create new industries, jobs, and wealth, a far more productive endeavor than the current obsession with management of scarcity. Space can provide new pharmaceuticals to cure diseases, personal communications for millions of citizens, nighttime lighting for our cities, limitless and non-polluting energy, a frontier for tourism, and a host of other services which are likely to be relevant in the daily lives of millions. These services will also enable the developing nations of the Third World to progress faster while leapfrogging entire generations of unnecessary technological evolution."

This broad overview is not the work of a science

fiction writer, but the considered expectations of one of our leading planners in space development. Furthermore, he emphasizes that nothing he is proposing requires any major scientific breakthrough. After outlining a whole series of developments which would extend and improve the quality of life for all Americans, he asserts:

"All of these possibilities can be realized via space. Most of the satellites are large and complex, and require assembly and checkout on a manned space station prior to being gently transported to their final orbit. Some will require cutting the cost of transporation to space by an order of magnitude, a goal possible by 2000 with rockets already on the drawing boards. And concepts are known that would reduce the cost by another magnitude in the further future. In fact, such transportation cost reduction will benefit all space users. Availability of cheap transportation has always been the key to progress and opening of new frontiers. So with space, load availability will allow space tourism, beginning at first with modest cabs for a few people but surely leading to space 'Hiltons.' Thus space will create a whole new industry beyond the year 2000, whose growth is expected to rival or exceed that of transportation in the early 1900s and communications in the 1950–80s."

Another key space planner at NASA argues that our society has no choice if it hopes to succeed and survive:

"In general a society that can recognize the future course of events and take a prudent lead in those events will remain vigorous and strong. If a society backs away or is hesitant or rests on past laurels, it will be left behind. A successful society at every moment must reach beyond itself because nothing stays the same. To remain strong and vigorous, a society must continually probe new ground. To be sure, a society cannot do this without deep concern for its problems. It must be just and fair, it must be economically sound, it must protect itself from dangers. But if it only does these things without challenging itself by continually forging new opportunities, it will ultimately wallow in itself."

The amazing fact was that America literally stood on the Moon and watched in its living rooms as the dream of freedom reached out beyond our planet in 1969 and yet we turned back and wallowed in the problems of the welfare state for a decade. Food stamps crowded out space shuttles; energy assistance crowded out a solar power satellite project that would have provided energy for all; more bureaucracy in Health and Human Services shoved aside a permanently-manned space station; the vision of a malaise-dominated decaying Western culture smothered the dream of a permanently-manned station.

Post-Watergate reaction turned people against space and starved the shuttle (thus driving up its cost by stretching out its development and procurement), and also ignored all the other major possibilities beyond our planet.

But, even as Jimmy Carter told us we were suffering from malaise, Americans were going to see *Star Wars.* Even as Teddy Kennedy attacked Carter for doing too little about poverty, people were going to see *Star Trek: The Motion Picture.* The American people wanted to rekindle the dream.

That is why they responded when President Reagan in his inaugural address reminded us that we "are too great a nation to limit ourselves to small dreams."

That is why they responded when President Reagan in his 1984 State of the Union Address declared, "Tonight I am directing NASA to develop a permanently-manned space station—and to do it within a decade."

The Reagan proposal for a permanently-manned space station is an important step in the right direction. While the initial planning will only cost $150 million in fiscal 1985 and $280 million in fiscal 1986, the station will ultimately cost about $8 billion for the initial version to be completed by 1992. The follow-on modules which will be added by the year 2000 will cost an additional $12 billion.

NASA's share of the space station will virtually all

come through a constant dollar budget (actually one percent growth per year), which transfers funds to the space station project as the space shuttle project winds down and commercial users of the space shuttle start paying for more and more of the operations and maintenance cost of the space transportation system.

The space station will provide opportunities for a variety of new commercial and scientific projects: the first stable base for long-term production of extremely pure drugs, of gallium arsenide crystals for use in microchips, of alloys developed by mixing metals (lead and aluminum) that won't mix in Earth's gravity, of extremely clean materials that need a near vacuum (which is very expensive on Earth), and of ultralight, very large structures that can be built with extremely thin materials in a zero-gravity environment but would be crushed by Earth's gravity.

The initial American space station will probably have four modules, each about 22 feet long by 14 feet wide, pressurized containers to provide working and living space for the crew. With a living space module, a logistics-storage module, and two laboratory modules, a permanent manned presence in space begins.

The initial space station will probably include a biology and health care module, a chemistry and manufacturing laboratory, and a service bay in which satellites could be repaired by crews wearing spacesuits, as well as a resources module to provide the energy and life-support systems for the four human-habitat modules.

This first permanent American station will probably have two platforms that fly independently of the station, one to serve as a combined astrophysics observatory and materials processing station, the other to fly a polar orbit for Earth-viewing sensors combining atmospheric, resource analysis, and defense-related functions. This second platform could also form the first step toward a potential polar-orbiting second or third manned station.

Such a permanently-manned space station provides a range of important opportunities for the advancement of science. Our understanding of the universe and the Sun, the study of planets and their environments, the study of the planet Earth, including the constant changes in our dynamic environment and our interaction with the Sun—all these areas will be better understood once we have a permanent platform for observation and analysis.

Environmentalists should be particularly interested in developing an Earth-observatory in space because our ability to monitor and keep track of pollutants, changes in topography, and changes in temperature will increase substantially with a permanent observatory. We will eventually develop monitoring systems for our entire planetary environment, just as today we have monitoring systems for a hospital bedroom or a critical laboratory and factory.

Industralists interested in crystals, semi-conductors, and optical fibers will have a laboratory in which to develop zero-gravity technologies beyond our current imagining. Projects to convert sunlight into laser and microwave energy may become not only feasible but also cost-effective, once we can undertake experiments free of the Earth's atmosphere. The development of large, lightweight space antennas may lead to dramatic breakthroughs in computer science and our capacity to handle information.

Our communications industry's capacity will increase dramatically once we are able to assemble very large communications satellites in space. Today, the problems of size and blastoff tolerances limit the power of our communications systems.

By the late 1990s, the demand for communications through satellite links will have exhausted the capacities of the current generation of satellites. New antennas far too large for the space shuttle to carry will be necessary. Furthermore, orbital transfer vehicles will be needed to move very large assembled satellites from the space station's low orbit into the much higher geosynchronous orbit.

The permanently-manned space station offers the United States more than merely commercial and intellectual possibilities. Properly developed, the manned station will tighten the bonds of the Western Alliance and provide a stepping stone toward more common effort with Third World countries.

President Reagan has already called for an effort to involve our immediate allies in the development of a manned space station. NASA is already exploring with the Canadians, the Japanese, and the Europeans the possibility of various programs which would share both the cost and the technological opportunities of moving mankind permanently into space.

The Canadians, who are interested in creating a second-generation manipulator arm to follow up on their success with the arm presently incorporated in the space shuttle, also have a vested interest in a polar-orbiting Earth resources platform which could help them analyze their huge country. The European Space Agency pioneered work on pressurized modules in the Spacelab effort; the development of a European module with German, French, and Italian participation is almost certain.

The Japanese, as one might expect, are focusing on the commercial opportunities inherent in materials processing. By the end of the century, there will probably be a Japanese manufacturing module, the first step toward an independent Japanese permanent space station.

Besides the participation by developed countries, the probable year of the space station's opening, 1992, creates a real opportunity. 1992 will be the 500th anniversary of Columbus' discovery of the New World. The space station represents for our children a new world just as Columbus' discoveries did for his era. The United States might increase its prestige and popularity in Latin America by consciously linking the station to the Latin American celebration of Columbus' achievements. Latin American nations should be invited, through the Organization of American States, to join in building and manning their module. By drawing in a Third World, non-military partnership for one of the modules, the United States could emphasize our common heritage and common bonds with the entire Western Hemisphere.

The United States will gain far more supporters and do more good throughout Latin America by including Latins in the space program than it will by sending the same amount of money to any Latin American country in military aid. The fact that we are willing and eager to assist Latin American scientists and Latin American engineers to take what Ben Bova calls the "High Road" would be appreciated throughout all of Latin America.

Furthermore, if we don't include Latin Americans in our space program, the Soviets will take advantage of our neglect and include them in theirs. It would be truly destabilizing, as well as embarrassing, if the Latin American perception at the end of the century was one of the Soviet concern to involve them in space and U.S. indifference to its neighbors. This possibility is very real.

The Soviet space program is at least as large as the American program. The American six- to eight-person, 80,000-pound permanent station has to be compared with the twelve-person, 300,000-pound space station which *Aviation Week & Space Technology* suggests that the Soviets will have before our station is finished.

The Soviets just keep plugging away: when we halted our aggressive program after Apollo, the Soviets continued to make steady, incremental developments. Their satellites and stations are not as elegant as ours, but they compensate by putting up more and bigger systems.

For a sense of the scope of Soviet effort, compare our astronauts' seven- to ten-day trips in the space shuttle with the record of Valery Ryumin, a Soviet Cosmonaut, who has made two voyages on Salyut 6 that each lasted six months: by himself, Ryumin exceeds the entire American man-days in space with the shuttle system.

Nor are the Soviets the only competitors who have been steadily working while we have marked time. The Japanese believe they can compete with us in communications and resource satellites and in expendable launch vehicles.

Expendable launch vehicles will handle at least 40 percent of the space launching business in the next generation because they are more convenient, for small satellites, than the shuttle, so the development of Japanese competition, when added to the European-built Ariane, poses a real threat to the American space industry. If both our Free World competitors succeed, the United States may lose a lucrative foreign market. Even now some American companies are using foreign rockets to launch satellites, taking advantage of favorable financial terms offered by the French.

The Japanese have intensive efforts under way in the form of both the McDonnell Douglas-Mitsubishi N-2 launch vehicle and the Toshiba-General Electric satellite-development program. By co-venturing with American companies, the Japanese are gaining the technical expertise and management insight to be very serious competitors in the next generation in space.

As a very job-oriented nation, the Japanese understand the second- and third-order effects of developments in space. This is why there is already a new industry building family-size antennas to receive television signals directly from space. If the United States coasts long enough, we may find a majority of the next generation's space-related jobs is being developed in Japan.

The fact is that, even with President Reagan's commitment to a permanent space station, the United States is still using only a tiny portion of its resources and potential to develop our future in space.

Consider just a few facts:

While the space program may look large in isolation, it is truly tiny when viewed in comparison with either the rest of the American government or the American economy: the NASA budget for an entire year would run the Defense Department for eleven days and equals the Health and Human Services budget for only eight days.

Even when measured against the American private economy, the space program is not massive.

Over thirty American corporations are larger than NASA. Indeed, many of NASA's contractors such as RCA, General Electric, IBM, Westinghouse, and Western Electric are bigger then NASA; General Motors and Exxon are both more than ten times NASA's size by themselves.

This misconception of the size of NASA's budget is due in part to the effect of inflation. On paper, the NASA budget is much higher in 1983—$6874 million—than it was in 1965 at $5250 million. However, the modern dollar has been savaged by inflation. When viewed from a constant-dollar standpoint, it becomes obvious that 1965 was clearly the peak year for real purchasing power at NASA. Today's NASA budget in constant dollars is only forty-three percent the size of the 1965 peak: in other words, the NASA 1983 budget of $6874 million would have to have been increased to around $16,039 million to have equalled the purchasing power of the 1965 peak NASA budget.

Thus, despite all the talk about "bold new steps," the reality is that President Reagan's *entire* new manned space station could be bought in a single year if we were operating at the 1965 level of intensity. Imagine the dynamism and jobs that would be created in America if we were building a new space station this year, three new shuttles next year, and creating a lunar base in the following two years—a pace that the 1965 budget would sustain.

If we have all this potential in research, in manufacturing, and in opening up new worlds for all mankind, why don't we do more?

If we have the basic technology necessary for enormous breakthroughs in space, why don't we do more?

If we have political leaders who at least *talk* about the dream and who understand the importance of space for both civilian and military use, why don't we do more?

The answer to our timidity is in part the bureaucratic nature of NASA and the lack of interest groups fighting for a bigger space budget. However, the underlying answer lies in the very cultural assumptions of a welfare state and in the decadent biases of the Liberal intellectual elite.

Dan Rather responded to the proposed space station by raising the right cultural questions in response to his own son:

On a November 28, 1983, radio commentary, Rather reported, "My son and I are talking about the space shuttle, and he shakes his head over my continued excitement.

" 'Pop,' he says, 'it's not a big story anymore. It's old hat.' "

Rather went on to comment, "I think we've forgotten that the notion of exploring space is something that makes people sit up and take notice, something every welder and waitress can get into. Space is the high frontier, the next frontier, the last frontier, and man loves to push out and discover.

"Today the space shuttle orbits the Earth, and that's a form of pushing out. But what about manned landings? What about the Moon again? And what about Mars?"

Rather concluded, "You wonder if perhaps—just perhaps—the U.S. ought to be paying a little more attention to the idea of a return to manned landings. Man is a searching animal and a learning animal, and how exciting it would be to land at another planet and discover.

"Do we have to wait until way after the turn of the century? I wonder if the question doesn't come down to this: Will we be like the Portuguese who didn't want to spend money, or like the English who profited greatly from pushing on and pushing out and setting foot on new worlds?"

That commentary raised the fundamental cultural issue which President Reagan's proposed manned space station only barely addressed.

The space station is a stepping stone, a focal point. It is the next step after the space shuttle. It is logical, bureaucratic, planned, and methodical. It can be more exciting than the shuttle. But if it is thoroughly smothered by the NASA style, it will be one more *Right Stuff* when what we need is *E.T.*

The welfare state is, at best, trapped in boredom because the "Two Cultures" of which C.P. Snow warned now exist. Snow, a British physicist-politician-novelist, warned that the Western world was increasingly splitting into opposing intellectual traditions: one which understood the humanities but was ignorant of science and mathematics; another which was immersed in the sciences but had relatively little knowledge of the humanities and social sciences. The result, Snow warned, would be a society in which those who shaped opinion and governed knew relatively little about the forces which changed their society, while those who commanded the forces of change would find it difficult to communicate with either political or cultural powers.

The United States began suffering an acute case of the Two Cultures syndrome with the Free Speech Movement at Berkeley. The Civil Rights protest movement, the anti-Vietnam War protestors, and the Watergate scandal added new impetus and energy to the crisis. By the late 1970s, America was caught up in an epidemic of technological abhorrence led by much of its intellectual, political, cultural, and news-media elite.

It is normal for the young to rebel. Had the American experience been limited to the Free Speech Movement, there would have been a brief surge of youths testing the will and morale of their elders. Then the young would have gone back to learning.

But the initial momentum of the Free Speech Movement was reinforced first by the moral righteousness of the civil rights demonstrations and then by the fervor of the anti-war movement. By 1968 and the riots in Chicago, an entire generation of activists had concluded that the values and systems of their parents were wrong.

The long agony of Vietnam was followed by the disgrace of Watergate, and the already suspicious activist-intellectuals felt vindicated: the system really was corrupt; the dangers of a high technology society were such that many young potential leaders opted for communes, hippydom, and emotional causes that demanded sign-carrying but not studying.

The anti-technology movement thus became a justification for not mastering the mathematics and science which are so crucial to America's future, as the Vietnam and Watergate episodes became justifications for avoiding the real problems of building large systems capable of achieving large goals. In a very real sense, our hippies overshadowed our astronauts and the anti-technology bias of the Left overshadowed the possibilities of the computer age.

As technically skilled people lost status, the news media found it unrewarding to cover the developments in high tech that would shape the future. TV's demands for sensationalism and quick answers made it almost impossible for physicists and chemists to explain themselves. A cooled-out generation of newsmen and women, more used to cocaine than computers as an after-dinner topic, found little of interest in the technical revolutions which were gradually changing the world.

Investigative reporting was the order of the day, and hard-hitting, tightly focused exposures of business and government wrongdoing, not serious explorations of technological change and the opportunities it would create for ordinary people, were the reporter's ticket to ratings and promotion.

As the '80s dawned, society slowly began to regain its balance. The hysteria of the early '70s began to give way to selected glimmers of hope: PBS's *Nova* television series introduced millions to new ideas; the development of magazines like the *Smithsonian, Natural History, Discover, Omni,* and others complemented the traditional role of *National Geographic* and *Scientific American* in educating readers; publications about home computers poured onto the market; a new generation, bored with the decay and narcissism of its elders, began to want more than dope and decadence.

Carter's "Malaise" speech in the summer of 1979 was the high-water mark of the anti-technology wave. Americans who listened to the President tell them that there was no hope began looking for ways to reinvent hope.

President Reagan's commitment to a permantly-manned space station is not enough in itself; neither is commercializing space, although it will help create jobs and generate a network of pro-space groups.

An opportunity society requires a dynamic, expanding vision of freedom and the future. If people are to forgo eating their seed corn, they must believe in a worthwhile harvest; if they see a vision powerful enough, they will commit themselves willingly to a thousand-mile march. Without that vision, they are unlikely to move at all.

The crisis of our Two Cultures has delayed an articulation of the nascent opportunity society. Those who know how to speak and write have been technophobic; those who understood technology's promise for the future have generally been inarticulate.

Today, a new generation of politically and culturally sophisticated writers is making a difference. Writers like Jerry Pournelle and Ben Bova understand that space has to be attractive to the general public if America is to move into the future. Greater general awareness has given society the means to plan its future.

There is a slowly growing understanding that we not only have a right to dream; we have an absolute need to dream. It is the power of dreams which gives us the energy and drive to work together for a better future. The absence of dreams leads to a nightmare of inward-looking decay.

When those who dream are ignorant of technology, they envision only a limited future of declining resources and rising pain. They proclaim a limit to growth and a need for massive bureaucracy to spread the misery equally. It is not their fault; they are simply ignorant.

However, if we combine our historical optimism, our free market drive, our capacity to use the government to build large systems (like the Panama Canal), and the technically feasible opportunities in space, we find more than enough romance and hope to motivate us for the next several generations.

Consider, then, five simple steps to a bold future:

1. Spend as much on civilian space as on defensive measures in space. This would both emphasize our civilian commitment to the rest of the world and obligate the military to consider subsequent civilian uses of the technology and hardware it develops because a generous civilian space budget will make the military and NASA natural allies.

2. Revert to NASA's 1965 peak budget in constant dollars. Make a national commitment to keep the budget at that level through the year 2000. With that level of resources, we will lead the human race into space.

3. Change the tax and regulatory systems to increase the speed with which we commercialize space. It is in our long-term interest as a nation to create a climate in America conducive to the exploration of space—the *only* frontier—which encourages the risk-taking and daring that only private initiative can risk. The longer we keep space a government monopoly, the greater the danger of smothering initiative.

4. Involve our allies and our Third World friends in the development of space. If we lead with a welcoming hand held out to others, the world will join us. Over the next generation, the combination of U.S. high technology, our openness, and our generousity can attract young Third World dreamers eager to work in space with Americans. In addition, it will tie Europeans and the Japanese to us in a way that no other inducement could match. Space development, a participatory process with the developed Third World, will increase our prestige more than trillions in foreign aid.

5. Develop the beginnings of populism in space and the beginnings of citizen involvement by initiating a tourism program for the American people: Americans who have paid the taxes deserve to enjoy some of the benefits. The shuttle is already comfortable enough to carry anyone free of severe health problems into space, and the next generation shuttle

will be even more like an airliner. We should begin with a candidate-selection lottery based on individual income tax forms and offer to send each year's winner on a shuttle flight. As missions become more common, we could send up a couple per flight. More than anything else, this would tie the average American into the space program.

The kind of dynamic, well financed, worldwide program outlined above could lead to more exciting developments like those discussed in June, 1979, when NASA convened a New Concepts Symposium at Woods Hole, Massachusetts, where thirty experts brainstormed a range of pioneering options for NASA worthy of Lewis and Clark. Had Carter listened to these intellectuals, instead of harkening to the defeatists, he might have been re-elected.

Consider just a few of their proposals:

The Astrophysics group thought we could build a multiple-mirror telescope 100 meters across. They also proposed a 1-kilometer diameter, ultra-thin plastic film lens to focus radio waves at a receiver 100 kilometers away in space. And, in their most daring proposal, the group suggested that it might be possible to develop a receiver to detect gravity waves. We cannot detect gravity waves on Earth because of background distortion, but their nature, existence, and behavior are crucial to our theoretical physicists. When I think back to Benjamin Franklin's experiments with lightning and electricity and all that led to in our world, I am inclined to believe that gravity waves will yield discoveries of practical significance.

The Climate group at the Woods Hole conference suggested that a large array of mirrors could affect the Earth's climate by increasing the amount of sunlight received by particular areas, citing recent feasibility studies exploring the possibilities of preventing frosts in Florida or enabling farmers in high altitudes to plant their wheat earlier.

A mirror system in space could provide the light equivalent of many full moons so that there would be no need for nighttime lighting of the highways.

Ambient light covering entire areas could reduce the current danger of criminals lurking in darkness. Mirrors could be arranged to light given metropolitan areas only during particular periods, so there would be darkness late at night for sleeping. They also proposed two 100-meter diameter radars to be focused on severe storm systems and analyze wind and moisture patterns. This type of detailed, constant monitoring of large storm systems would increase our understanding of some of the most destructive and dangerous natural phenomena—hurricanes, typhoons, and blizzards.

In the Communications group, some planners proposed a wrist radio system that would provide tracking capability for people and some limited communications capacity at a relatively low cost once the basic system was installed. There is no technical problem in building such a system, though there is a very lively debate over whether or not it would ever be economically feasible. The Communications group also emphasized the long-term need for more communications and information capacity than our current satellite systems permit. Many low-altitude satellites, needed for ground-links, could be linked by laser to geosynchronous satellites in more distant orbits for long-distance relays.

The Energy group brought into focus one of the major problems in bureaucratic modeling and planning: the NASA-Department of Energy study of the solar power satellite concept had focused on a relatively expensive and heavy model that, given current transportation costs, is not competitive with nuclear power and would be a poor investment. However, the Energy group considered a number of alternative designs which use the nature of space to permit large-scale structures at much lower cost. One solar power satellite version they proposed would have panels ½-kilometer wide by 2-kilometers long. A total of twenty-four panels would connect on a 70-kilometer long satellite. The Energy group estimated that this particular system would weigh far less than the NASA-DOE model. As a result, it would cost

30-40 percent less than the solar power satellites NASA studied.

An additional factor in developing less expensive solar electric energy is the declining cost of photovoltaic cells. At the present time, solar (photovoltaic) cells tend to convert sunlight into electricity at around 12-14 percent efficiency. I have talked with one entrepreneur who believes his company will shortly be in the 50 percent efficiency range. Such a breakthrough would drop the cost of solar power satellites by the same factor as the efficiency increases.

If we project a one- or two-magnitude drop in space transportation costs by the next century and a use of robotics for self-replicating systems in space (see below), as well as an increase in the efficiency of photovoltaic cells, we can expect solar power satellites at fully competitive costs early in the next century.

The real gateway to space has been the development of steadily less expensive methods of boosting payloads into orbit. However, while the space shuttle is a major step forward, it still costs in excess of $1,000 per kilogram to fly cargo to low Earth orbit. The same weight could be flown from New York to Los Angeles today for about $5. According to the Woods Hole conference's Transportation group, that discrepancy exists despite the fact that the same amount of energy is needed for both trips. On an even more striking scale, the same quantity of energy in the form of electricity costs about fifty cents.

We can make major improvements in the transportation system used to put cargoes into space. Some experts think that the second-generation space shuttle should provide a one order-of-magnitude improvement, so that it would cost about $100 to put a kilogram in orbit. Then the third-generation shuttle of the year 2020 should offer yet another magnitude drop about $10 a kilogram. At that point, a typical couple might take a honeymoon trip into space for around $15,000. Passengers are likely to be subsidized at first by cargo, just as air mail subsidized virtually all the early American passenger airlines.

Tourists may well reach space before they are economically profitable in their own right.

The Transportation group looked at a number of fascinating alternatives to the traditional chemical engines we are familiar with: laser-based power systems; ultra-thin solar sails; tethers 200-kilometers long to winch payloads from a suborbital shuttle to a low orbit platform; and a flipping system of tethers to toss cargo from low orbit into a geosynchronous orbit. This variety of ingenious approaches may be worth exploring as we seek to lower the cost of transportation.

As we plan for the next generation in space, it is vital to remember that we are at a stage comparable to the early days of aviation. The pioneers who carried one, two, and three people in biplanes for very short distances had dreams of a Ford Trimotor, but no inkling of a Boeing 747. Pioneers someday hoped to have a technology which caught up with their imagination; reality surpassed imagination. Through a great deal of government subsidy and encouragement, including the Civil Aviation Administration, airport construction, development of navigation systems, subsidization of the air mail, research and development grants of the U.S. Navy and the Army Air Corps, and favorable tax and regulatory policies, the United States gradually developed a massive export industry and the finest civil aviation industry in the world.

Today, we who believe in space can envision our Ford Trimotor: we are trying to encourage planners who look at 2010 to envision at least a one-magnitude drop in transportation cost as the second-generation shuttle takes over. It is vital that we dream expansively if we hope to interest high school students of today in becoming the physicists and engineers of 2010.

Another breakthrough that will make space profitable will be self-replication systems. Von Neumann originally outlined the concept of systems which would reproduce themselves over and over so they could develop exponential growth. Today, the technology of information systems and robotics will soon permit self-replicating technology to advance our capacity to manufacture in space.

Self-replicating systems begin with an initial system which then builds a duplicate of itself. Once it has finished, both it and its duplicate go to work building duplicates of themselves. The power of such exponential growth is awesome. A self-replicating machine could rapidly build a factory or a lunar base or a large system in space. The key to such a system—providing the system with enough materials at low enough cost—is the lunar bases.

The Moon is an enormous natural resource, possessed of more than enough minerals and materials to provide everything a self-replicating system needs. Structural glasses and ceramics can be made by crushing rocks and molding them by heat; oxygen and water can be manufactured from the Moon's soil to form life-support systems for humans; iron and titanium are abundant on the Moon and can be extracted for use.

The first self-replicating systems might be designed to build a variety of materials-processing factories, then a series of materials finishing factories. Presently, photovoltaic cells could be built on the Moon using lunar materials. Since the Moon has only one-sixth the gravity of the Earth, the cost of placing a lunar-made solar cell in Earth orbit would be less than that of bringing the same material up from the Earth.

The NASA study group suggested that an initial system could multiply itself rapidly enough to produce 1,000,000 pounds of solar cells in no more than twenty years.

Any large investment in machinery requires humans to service the machines and to respond to the unanticipated. Thus, a significant commitment to a self-replicating system will mean a permanent manned lunar base. Indeed, the bigger the investment in robotics and manufacturing, the more certain it is that people must be on-site to repair the unexpected and the unavoidable.

At present there is a fight between the planetary

scientists who debunk a manned station and the manned-space advocates who debunk robots. Both sides miss the point. We want both people and machines in space, in large numbers, as rapidly as possible. We want machines to do the many things machines can do; we want people up there doing those things only people can do.

The infighting between planetary and other scientists on the one hand and pro-manned space advocates on the other has been peculiarly shortsighted and debilitating. Each must reinforce and help the other so both can win. The secret to success in American history has been to find enough partners who share a vision to bake a bigger pie for everyone. That bigger pie must include a permanent lunar base and we should plan the founding of that base to coincide with the coming of the next millennium.

A year which ends in three zeroes is a rare thing indeed. As the year 2000 approaches, more and more people will notice that they are about to celebrate something which no ancestor for nearly thirty generations saw, and which none of their successors for another thirty generations will see.

How then shall we celebrate a millennium's change?

The principles of celebration are simple: we should do something in concert with all the other free people of the world to show that our joint commitment to freedom rises above nationalism; we should do something which celebrates the power of high technology that will remind us and everyone else that the greatest single factor in the rising standard of living over the last millennium was not our politicians and academic intellectuals, but rather our inventors and business entrepreneurs; we should do something which holds out an improving future to the entire Third World so that everyone can realize that our path, rather than Castro's dictatorship, is the wave of the future; finally, we should do something which is peaceful and knowledge-oriented as a first step toward creating a Human Peace in the next millennium.

The most appropriate single millennium project

would be the opening in January 1, 2000, of a lunar research base for the whole free world.

The early, permanent lunar base would house only four to six people, probably on a three- or six-month rotating cycle much like the regime in Antartica. In the early days, the two primary jobs of this base would be establishing the self-replicating systems and undertaking initial scientific observations to establish long-term base lines for research.

As self-replicating systems developed, the lunar base would become self-sufficient in water and oxygen. Over time, enough soil would be formed to provide for greenhouse agriculture in controlled environments. Additional self-replicating systems would be added, first to expand the size of the base and then to build new materials for use in space. By 2050, most of the semi-finished material used in space manufacturing could be of Moon origin rather than Earth origin.

The first decade on the lunar base will be austere and difficult, but with the exponential growth capacity of self-replicating systems, the lunar base might be a real colony by 2010 or so. One senior space planner estimates that, by then, the colony could support one-hundred to three-hundred people. Follow-on growth would be even more rapid. By 2050, the population of what would by then be several lunar colonies could be in excess of twenty-five hundred people. Their robotics-supported manufacturing and agricultural systems might be close to self-sufficiency.

If all this is possible, why are there so many doomsayers preaching the limits to growth and arguing that we face a desperate and dangerous future?

The reason is political and cultural. The welfare state could have built a permanent lunar base in 1969—it decided not to. The welfare state could have raised a generation of committed young mathematicians, scientists, and engineers eager and excited about going into space—it decided not to. The welfare state could have reached out to our allies and to the Third World and built a cooperative venture that would have knit all freedom-loving people together in building a better future for all mankind—it decided not to.

Today we stand on the verge of the movement of a "Unified Free World Alliance" into space. We stand on the verge of a grand commitment to a brilliant future. We stand on the verge of an investment that will liberate our grandchildren. As Jim Muncy, President of Using Space for America, wrote in his essay "Needed: A Bold Leap into Space":

"If the space program is to be the vanguard of our nation—and well it should be—then a bold nation demands a bold space program. And therefore our goals in space must by definition be visionary enough to set real challenges for ourselves. A space program that focuses on tactical steps of hardware will inevitably force the public to think in terms of costs vs. benefits. If instead the space-program aims toward long-term goals, it will reinforce our determination to invest in a better future of prosperity and freedom. . . . It's time we have the courage to dream a great dream again, and set a real goal for America in space: a permanent manned lunar base."

That, then, is the scale against which to measure our dreams. We must not be penny wise and dollar foolish, setting a small budget and losing the opportunity to shape the future of all mankind. We must have the courage to do as our ancestors did: to take fragile vessels on dangerous voyages to unknown lands.

As we answer this challenge, so shall our descendants know who we are. As Americans, I truly believe we can only give one answer. The Adventure begins—

CHAPTER 3: THE INFORMATION EXPLOSION AND THE GREAT TRANSFORMATION

The most powerful force changing our society is the information revolution. It is as powerful as the word "revolution" suggests, but it is a more complex phenomenon than popular analysis describes.

The great force changing our world is a synergism of essentially six parts:

1. Computers.
2. Cable and telephone wiring of our communities.
3. Satellite distribution for global and long distance transmission.
4. Television and radio broadcasting.
5. A knowledge base which has been growing for thousands of years and is now doubling every few years.
6. The miniaturization of components and systems.

The information revolution is based on the interaction of these six factors. No single development would have the power to transform our world. It is the exponential interaction of all these factors which has led to an extraordinary growth in knowledge and in speed of communication and information processing.

The closest analogue we have to the information revolution's potential for changing our lives is the impact the utilization of electricity had on our great-grandparents. Electricity, like information, is a source of power rather than an end product. Electricity transformed everything with second- and third-order effects, many of which became evident in the last thirty years.

The electrical revolution took shape over a century, beginning with Franklin's experiments with kites and lightning. It required laboratory scientists like Faraday and inventors like Morse and Edison. Many of the breakthroughs involved small amounts of electricity—the telegraph, for example; other inventions, such as the electric railroad and the processing of aluminum, required enormous quantities of electricity.

Any contemporary attempt by politicians and the public to understand the electrical revolution would have been frustrating and incomplete. In the heyday of its development, electricity was an unknown quantity for which new applications were constantly being developed. While there were famous developers of electrical inventions like Edison, Steinmetz, Siemens, and Marconi, hundreds—indeed, thousands—of other scientists and technicians at General Electric, Bell Labs, Westinghouse, and elsewhere made significant contributions.

Electricity changed our lives in ways none could have imagined in the 1840s: the icebox gave way to the refrigerator; the gas lamp was replaced by the electric light; the servant with a fan was replaced by air conditioning.

On a larger scale, electricity changed the world: refrigerated railroad cars and steamships permitted the growth of an agricultural market unthinkable prior to electricity; the telephone allowed people to communicate conveniently over distance on a scale unimaginable prior to electricity; the motion picture and then radio and television changed how we relax, how we learn, and how we live.

Today we have some thirty-five electric engines per home and we take all of them for granted. They don't normally require maintenance or any other attention—plug them in; turn them on.

Today, we take the great revolution of the late 19th century for granted. No one marvels at the telephone, the electric light, or the motion picture. Instead, we miss them when we go camping. We have grown so accustomed to our modern conveniences that we now regard the manufactured world as natural.

The information revolution can be best understood in the context of the electrical revolution. Focusing on one aspect of the information revolution, such as home computers, would be like focusing on washing machines in 1880. It is true that home computers are marvelous and exciting, just as the first washing machines were exciting, but home computers are a very thin part of the revolution we are now experiencing.

In a sense, we are in the pre-Edison days of our information explosion. Edison, prolific and extraordinarily disciplined in his approach, undertook over four thousand experiments in order to develop a practical light bulb.

Edison's impact came not simply from his genius or intensity, but from his determination to make technology available to individuals as well as society. Edison was the inventor for every man: he understood that technology, to be useful, has to be usable; if the use is beyond the comprehension of the average individual, it will remain out of reach in practical terms.

The information revolution has lacked a genius thus dedicated to making information everyone's servant. There have been many geniuses involved in the development of computer chips and software, many brilliant teams solving the technological problems, but relatively few people or companies working from the consumer back to the computer. The primitive nature of the information industry's orientation to customers is obvious in any course on computer use: even the most basic courses focus far too much on the theory and construction of computers rather than on their uses.

The trend of technology has been that machines change to fit people far more than people change to fit machines. Early automobiles were essentially mechanics' playthings, requiring a wide range of mechanical understanding, beginning with the ability to use a starter crank for igniting the engine. Today's automobiles have switches and idiot lights: for a century, the automobile industry has been reshaping the car to fit the driver.

Similarly, the electrical revolution has been one

long process of changing tools to fit people. As Professor Pete Jensen of Georgia Tech reminds us, no one needs to understand Ohm's Law to use a blender. If we designed household appliance manuals or courses the way we currently run computer classes, we would teach people a lot of basics about electricity before allowing them to turn on their electric razors or waffle irons. In fact, we now take electrical appliances so much for granted that we start playing with them without reading the instruction books. When was the last time you read an instruction book before using an appliance? That is how completely Edison and his successors have transformed electricity.

Seen in that context of easy access and immediate applicability, the information revolution is still in a very primitive stage. Information technology is at a level comparable to that of electricity when Samuel Morse invented the telegraph. The telegraph was initially a crude enough system and the mastery of the Morse code was so specialized a process that the combination bears some resemblance to learning to program a present-day computer.

The fact that we have so many engineers and mathematicians involved in computer development has made it harder to find an Edison to popularize this technological revolution. Once someone has made the personal investment of time, energy, and money needed to master the current level of information science, that person has a vested interest in solving the problems of his or her peers. Furthermore, specialists in information sciences do not notice or compensate for the enormous barriers between average citizens and the information sciences.

A parallel to this problem may be found in the transition from early systems of writing to the simplified Phoenician alphabet. Early forms of writing, such as Egyptian hieroglyphics, Assyrian cuneiform, and Chinese characters (the last still in use), required the memorization of a multitude of word and syllabic signs decipherable by only priests and scribes.

The Phoenicians solved this problem by inventing a very simple alphabet. By making the memorization of a few symbols very easy, they democratized learning so that almost anyone could learn how to read and write. The Phoenician alphabet had as much power of expression as the more complex and numerous symbols used by earlier languages, but the simplicity of the alphabet allowed many more people to master it.

At some point, the computer will lose its current aura of magic and become more like the telephone and the television, a useful but easy-to-master tool of human improvement. At that stage it will require only functional, rather than theoretical, understanding. All the talk about "What will we do with people too dumb to use the computer?" will end. Virtually anyone can answer the telephone, yet the telephone was a greater leap forward in technology than the computer. In our grandchildren's world, if you are smart enough to operate the telephone and the television, you will be smart enough to operate the computer.

What is needed, as Pete Jensen suggests, is the development of information appliances as simple and easy to use as household appliances. When that develops, we really *will* have entered the information age.

However, focusing on the computer and how we interact with it limits our vision too narrowly. The computer, though important, is but the tip of the iceberg of change. Today's computers—even the largest—are closer to the individual generating systems of the early electrical era rather than to the great central power stations which have become the hallmark of modern societies.

When Edison was developing a range of usable tools, his bankers and investors wanted him to build only small generators for factories and specific buildings. They believed the market world be more profitable if every company and every building had to buy its own generator.

Edison insisted that the future of electricity required that it be easily accessible. He subscribed to the great Jeffersonian ideal that a free society could only grow as strong as individual free men and women grew. If generators were restricted to the rich, then the fruits of advanced technology would be denied to average persons. Edison's fight for central generating stations was in one sense a fight to democratize electricity and make it available to everyone.

We require a Thomas Edison, a Theodore Vail, or a Henry Ford who has the grand vision of bringing information technology to everyone. Edison envisioned an America in which ordinary people used electricity to improve their lives, and produced inventions to make his vision a reality. Vail had a vision of people universally using the telephone, so he kept inventing management systems and corporate structures like Bell Labs until every American had a telephone. Ford thought everyone should have a chance to own and drive an automobile, and he made the first gigantic breakthroughs toward that dream.

The key thing to note is that in each case these industrial-technological pioneers had an overarching vision of the future which gave each of them a framework within which to focus their efforts. This made their lives and their corporations remarkably productive and useful.

One of the major problems with information sciences today is that everyone is so busy with immediate inventions and with the possibilities inherent in the immediate technology that there are surprisingly few people dreaming on a grand scale. The visionaries of the post-World War II generation are now seeing their dreams come true. Almost everything Von Neumann and his contemporaries thought could be done is being done. The progress has been faster and more dramatic than anyone would have thought possible in 1950.

The current generation of computer scientists spend their time keeping up with the newest developments and explaining them to their clients, their colleagues, and their students. The result is a system of thought in which each day's tactical triumphs become major accomplishments.

Much of the current revolution is breathtaking.

Gordon Bell of Digital Equipment estimates that the computer power available per dollar has doubled every year since 1962. Today's computer is four million times as powerful per dollar of cost as its 1962 predecessor. *Byte* magazine has grown incredibly in size and complexity; *PC World* magazine, dedicated exclusively to the IBM personal computer, is now so thick and profitable that it is shifting from monthly to bi-weekly publication. The breakthroughs of Steve Jobs at Apple, Nolan Bushnell at Atari, and others like them are beginning to change the whole approach to personal computers. The explosion in computer power and speed has been matched by the collapse of computer costs. There has been a flood of software. America and Japan are racing to develop the fifth generation of superpowerful computers. These developments are individually striking. Taken together, they form a kaleidoscope that is difficult to develop into a coherent picture.

Yet it is by sweeping dreams that societies shape themselves. The great difference between the American and the European approach to the great revolutions in electricity, internal-combustion engines, and chemistry in the late 19th century was the fact that Americans, who had a vision of how these technologies would transform society, developed technology for the popular market. By contrast, the European societies were much more isolated from the great technological breakthroughs. In Europe, the new technologies remained much more the playthings of the rich and the technical elite. The result was a loss of democratization, a loss of living standards, a loss of capital investment and job-creating opportunities, from which Europe has never fully recovered.

Today, we have an opportunity to shape technology once more so that our grandchildren can achieve the same quantum leap in living standards from the information revolution that our grandparents received from electricity and internal combustion.

One key to envisioning the future is to recognize the powerful impact of three factors: 1) embedded chips to make our tools smarter and easier to use; 2) second- and third-order effects of breakthroughs; 3) changes in the way we think as a consequence of the revolution in information.

The embedded chip is the only one of these factors which is essentially a hardware development. It is an important breakthrough because it will change most of the things we now take for granted. Already we can buy cars that talk to us about key functions, smart microwave ovens, smart alarm clocks. This is just the beginning. Interestingly, the revolution will occur more slowly in health care and education than it will in consumer goods. The relative capital investment is greater in our kitchens than in our children's classrooms or the patient-oriented part of health care. The teaching and healing professions will be relatively indifferent or resistant to key technological changes in daily activities. There will be language processors in cars long before they are in language learning programs. Cars will monitor themselves thoroughly before individuals have devices to monitor themselves without professional help.

The embedded chip will provide watch-machines to prevent burglaries and heat-sensors to detect forest fires. It will give us control over our immediate environment and an ease of achieving many sensory and physical objectives that would be as unthinkable to us today as an air-conditioned tractor would have been to our great-grandparents plowing behind a mule.

The second great factor in the information revolution is second- and third-order effects. Most planning is linear and too narrowly focused. As a consequence, it misses secondary effects which are outside the arena of the primary change.

The railroad train is an example of second-order change based on the original stationary steam engine. The Newcomen engine was designed to pump water out of coal mines so that coal could be dug from deeper seams which had been flooded. Originally the steam engine was so heavy and so inefficient that it was impossible as a transportation system.

Concurrent with the development of lighter and

more efficient steam engines, there was an experiment under way in the use of rails to allow horse-drawn coal wagons to carry more weight with less effort.

The combination of steam engine development with the totally unrelated railroad (meaning literally a road whose bed was made of rails) led to the locomotive and the modern train.

In 1969 television broadcasting, home television sets, satellite-based relays, and the Apollo Moon landing combined to allow each of us to see the great event in our living room as it happened.

Because second- and third-order effects are hard to foresee, no one in the mid-18th century deduced that the transcontinental railroad would develop from the stationary steam engine. No science fiction writer postulated that people would see the first lunar landing at home on television. Yet it is in the area of second- and third-order changes that technological revolutions truly transform societies. It is when we begin to liberate ourselves from old assumptions (only horses draw coal wagons) that change really begins to gain momentum.

Here are a handful of ideas which serve to illustrate the way in which a number of needs come together with a number of different technologies to provide for real changes in the way we live:

1. A home video-computer system which would film your golf swing, compare it with the idealized models, tell you precisely what changes to make, and then show you the appropriate computer-generated imagery of the right swing or the right stance. This is, at most, twenty years away. It will combine with the emerging videotape lending libraries, so that your computer will tell you which tapes to borrow and watch to learn the right swing.

2. A personalized health chair with a diagnostic program to measure and compare all your bodily signs against your own data base. The chair could be tied into a weight-watcher's computer-based recipe program which would then outline what you should eat given your weight, blood-pressure, etc. The com-

puter could be programmed to monitor your diet over time and change recipes to minimize boredom while achieving the desired nutritional effect. This system could be tied by cable or telephone to a hospital, where a computer could routinely monitor you while you were sitting in the chair. Thus, you could leave the hospital after surgery much earlier than we currently expect; you could measure your own well being and take corrective and preventive health care steps; and you could monitor your diet and exercise patterns.

Such health chairs, which could be built with existing technology, would provide a higher awareness of personal health for the entire family and allow a lot more people to stay out of nursing homes.

3. Access to an interactive computerized income tax package and retirement rules and regulations package. This would permit all citizens to access rules required for their taxes and for their retirement accounts.

4. A computer directory and registration system for Federal parks, monuments, and recreation facilities. The Federal government offers every citizen literally thousands of vacation, recreation, and education opportunities. Many of them are ignored. In Macon, Georgia, the Ocmulgee Indian Mounds Park has a splendid natural walk area, a beautiful collection of ancient Indian ceremonial mounds, and fine museum on the history of the area from 900 AD to the present. Most citizens of Macon have never visited this park, though it is right in the heart of their city.

People complain about the cost of living and the difficulty of finding things for themselves and their children to do, when, in fact, there are scores of free opportunities. However, today you have to go to each separate government agency and ask the right questions. Since most people do not know which agency does what or what their addresses are, the information is too expensive in time and energy to be worth the effort. The information must be reorganized for the convenience of the citizens instead of for the bureaucracies.

5. An all-out effort to help people who happen to have handicaps or disabilities. The gap between what could be done and what we are doing is large and inexcusable. Every day the newspapers and magazines contain individual stories of people who are doing dramatically better because of an isolated breakthrough or a pioneering researcher. Yet there is no coherent national strategy to accelerate the development of programs to help all those who have handicaps or disabilities.

It is now possible for some patients with spinal cord injuries to use computers and electrodes to restimulate the muscles and regain the ability to walk. Some voiceless people have been equipped with voice-synthesizing personal computers.

Braille keyboards and Braille printers are major steps toward equipping the blind, but even greater strides are on the horizon. Combinations of radar and computers will allow the blind to sense and come remarkably close to seeing. This new technology provides the sensory equivalent of artificial sight.

Quadriplegics can blow on miniature sails to control switches. A cerebral palsy victim can work a computer keyboard using a pencil in his or her mouth. Stroke patients have regained their jobs through patient work on a computer screen. The prospects for liberating people who have handicaps or disabilities are growing with each month.

There has been no greater misdirection of effort in recent Washington budget fights than in the dispute between traditional Liberals who want to spend more helping Americans with disabilities and handicaps survive on welfare and those traditional conservatives who work to make sure no one cheats and gets on disability to loaf.

We should think in terms of a twenty-year budget for Americans with disabilities and handicaps. Capital investments in the first five years would lead to new technology that would allow the overwhelming majority of those with disabilities and handicaps to live fuller and more productive lives. Everyone I know who is in a wheelchair is longing to be more productive. All those I know who are blind are eager to do more for themselves and their communities.

The great challenge to us is to develop a vision of a completely participatory America in which everyone has a real chance to pursue the American dream. The revolution in communications and information can extend that opportunity to virtually every American if we will but take advantage of it.

6. A revision of the learning process to take advantage of new technology. Information is now increasingly electronic and available at the user's convenience and interest, but we continue to behave as though we lived in the age of books or even in the age of orally imparted knowledge. College lecture courses continue the medieval tradition of dictating to students who could not afford to buy what were then hand-reproduced books.

Libraries are still primarily the repositories of printed material. Some libraries are taking a half-step into the information age by buying computers for use in the library. This is an important first step and is not to be dismissed.

However, the salient reality of the information age is that knowledge will be stored, shipped, and processed electronically; that it will exist in such vast and rapidly changing quantities that people will seek to avoid all information except what they need to know and what amuses them; and that it will be at their fingertips through electronic systems.

The French are the pioneers in this aspect of democratizing the information revolution. The French effort to put a terminal in every house through the telephone system is an investment in the future and one which may make France the leading information-processing society in the world by the end of the century. The simple act of developing an on-line information network for telephone numbers and for the yellow pages instead of printing the telephone directory will both save money and permit a more fluid society by the end of the century.

The United States could also begin a series of experiments in home-based information systems using either the telephone or cable. The trick would be to encourage a lot of competing systems because we don't know enough right now to pick one mode over another. We need dozens of government-encouraged and subsidized efforts to build the information system of the future. Ultimately, all citizens will be able to reach the equivalent of the Library of Congress through their telephones or television sets.

The libraries of the future are available today, at least in pilot projects. A focus on how to package information electronically and how to get every rural and slum high school on line with the best information base in the world would do more to stimulate and renew education than all the current debate over how to marginally improve the teaching techniques of the 19th century.

7. A drive for adult re-education can be made through a combination of software, home computers, videotape lending libraries, and public television. We must do vastly more adult re-education in the next generation than we will ever be able to finance through the traditional methods of schooling. These new technologies permit self-instruction and self-help courses that start with the basics and move on up to very sophisticated programming. Ask any major corporation how many kinds of self-instruction and programmed learning it uses. Contrast that with the adult education programs of most states and localities.

Television, in particular, needs to be used to reach the very poor and the very elderly. We have to provide constant opportunities for learning so that no one is blocked from the information revolution that is changing our world.

8. Make the Federal government a showcase of the information revolution. The amount of capital investment per government civil servant is pitiful compared to the investment for a typical farmer, a factory

worker, or even an office worker in one of the more advanced corporations.

The jump from an electric typewriter to the interactive computer systems now on the market is fully as great as the jump from the quill pen to the typewriter. Government bureaucracy is largely a paper-processing operation. The more rapidly we automate and invest in a modern future for our civil service, the less expensive, more efficient, and more accurate our government will be.

We are also facing an increasing crisis in the morale and competence of our civil service. As workers see their office equipment and prestige decay, they lose self-respect and professional pride. Budget crunches lead to smaller or even no pay raises; pension reform takes away some of their past benefits; and political speeches denigrate their worth. Morale and daily performance suffer accordingly.

A decade of affirmative action has produced office after office in which people have their jobs because of quotas rather than competence. The office politics of recruitment and retention have now replaced the 19th-century office politics of spoils and patronage.

The result of all this decline has been a drop in the effectiveness of our civil service that is one of the understated crises in American society. In one recent town hall meeting I was told that a major Federal government office in Puerto Rico now has no one, repeat no one, on its staff capable of typing a letter. All typing in that particular department is now shipped to Atlanta to be typed. The Atlanta office is now having problems because it has barely enough typists left to do its own work. The competent ones are leaving to find less stressful, higher prestige, more rewarding jobs.

We could have a prouder and more efficient government if we made our Federal government a model of information and communications for the rest of the country. Part of that process would be capital investment in new equipment.

Administrators Gerald P. Carmen and Frank J. Carr have initiated a revolution at the General Services

Administration by opening the government's first computer store. Their initiative should be matched by the Congress in repealing the Brooks Act. That law ties government procurement up in so much red tape that it is currently impossible to fill out the paperwork and process the request before the proposed computer is made obsolete by new developments. No law with that effect can be allowed to stand unreformed. A great nation cannot remain great if its central governing agencies are demoralized, incompetent, and ineffective. Only a program like this can reverse that path.

The information revolution is also going to change the way we see the world.

Information technology is the result of bringing together communications, computers, software, and human information processing. The great breakthroughs of the last two generations were in the hardware of communications and computers. We are currently focusing more and more attention on software. The great achievements of the next generation may come in the field of human information processing.

How we as people interact with our new tools is an area of pioneering that has been relatively ignored until now. To put this field in perspective, consider how our view of industrial tools evolved.

The industrial revolution did not just trade us steam energy for muscle power. It reshaped our perceptions and assumptions about reality. We now live in a world of twenty-four-hour electric light. We travel over a global village by airplane and drive across our national village by car. Telephones allow us to talk at extraordinary distances.

The natural world will become more valued. An organic part of us will pull us toward nature even as technology permits us to avoid the natural world.

Our agricultural surpluses will be matched and then surpassed by industrial surpluses. Just as we produce more wheat than we can eat, we will presently produce more manufactured goods than we can use.

The focus of the industrial age was on the acquisition of *things*. In the future, the focus will be on acquiring knowledge and experience. People will only buy so many sets of furniture and so many cars before they start buying trips to Epcot Center or Paris or a professional athletic event.

We must remember that the *real* changes are the changes within us, not changes in the tools we use or the machines we employ.

The information revolution will raise the quality of life for all of us, and it will allow the entire Third World to participate in a richer and more exciting world.

CHAPTER 4: THE FUTURE OF THE WELFARE STATE

It is not hard to develop a pessimistic view of America's future. A handful of assumptions can wipe out all the potential progress and excitement developed in the earlier optimistic view of our children's and grandchildren's world.

The first step toward a pessimistic future would be for the national elite to reject the opportunities of high technology and the challenges of foreign competition and, by so doing, close the window of opportunity, perhaps forever: if protectionism ever becomes policy, the opportunity society cannot prevail.

The basic assumption that we can avoid change by building high tariff walls is simply wrong. A friend of mine has talked about writing a science fiction story in which the Japanese fleet comes to San Francisco harbor in 2054 and demands that we open up America. It is, of course, a take-off on Admiral Perry and the American navy demanding that the Tokugawa Shogunate open up Japan in 1854. Japan, in the early 17th century, limited contact with the Western world and, as a consequence, had fallen behind the West in industrial, economic, and military power.

Once Japan was forced to open up, the Meiji Restoration led a new generation of leaders to engage in a systematic effort to modernize their country, an effort which in many ways continues to this day, based on a systematic study of the West and the importation of those ideas which could be blended into the Japanese culture and lifestyle.

We stand at a crossroads: a decision to import and develop ideas that work and blend them into the American way of life will lead us to the optimistic future already outlined; a decision to seal us off will cripple America as surely as it crippled Japan for two centuries.

Read enough Liberal speeches and enough union brochures and it becomes obvious that there are many people in America who prefer isolation to alteration. There are elements—not solely Liberals—in our society that would blame America's economic problems purely on the unfair trade practices of other countries. Those who argue the most loudly for protectionism suggest that American workers have little to learn from the rest of the world, that American management has done a fine job, and that high trade barriers will allow us to avoid the need for change.

There are times and places when specific protectionist steps are appropriate: protectionism can defend an industry vital to national defense, can buy time for an industry to make adjustments to a sudden change in its environment, and can bludgeon a trading partner to force it to engage in fair trade. However, in each case the protectionist step should be taken purely as a bargaining tool or to enable American industry and workers to adapt so that we can regain a competitive edge—to make a transition toward a more competitive America, not to make a retreat into a fortress America surrounded by high tariff walls and quotas.

In the long run, a Liberalism that could not ideologically accept the challenge of high technology and could not educate its big labor, special-interest allies into changing into leaner, more competitive systems would be a Liberalism that led America away from the future toward a stagnant economic and technological system that rapidly would become non-competitive.

The problem with the first steps toward protectionism and the avoidance of change in a changing world is that those steps rapidly lead to even more decay. The less we change, the more we fall behind the countries which are changing constantly. Even the Japanese are finding it hard to keep up with their competitors in Taiwan, Hong Kong, South Korea, and Singapore. South America is becoming more competitive as well. The turbo charged engine of the Ford Thunderbird assembled in my congressional district in Georgia comes from Brazil.

Virtually every country is developing some kind of competitive industry, and there is increasingly a world market in goods and services. As more and more multinational companies share their production among a variety of countries, it is impossible to determine where a product comes from simply by the country of assembly. One estimate is that 60 percent of the European Airbus is made in the United States. That means that the first successful foreign competitor to American commercial jet airliner manufacturers is actually largely American. Conversely, Boeing is making parts for its new 757 and 767 airliners in Japan.

Thus, it is getting harder and harder to talk about protectionism in a world in which virtually every free nation is cooperatively manufacturing goods and services. Now more than ever, trying to protect domestic jobs by raising high tariff walls and quotas may actually kill more jobs as a result of the reaction of other countries, which will respond with tariffs and quotas of their own. Furthermore, a successful effort to force American companies to keep jobs at home would rapidly cause those companies to fall behind others laboring under no such restrictions. No matter how loudly American Liberals complain, the multinationals will share new production systems and technologies among the still open societies of Japan, Europe, and advancing Third World countries.

In many ways, multinational corporations are becoming the advanced economic training centers of the modern world. To the degree that we encourage multinational development in the United States, we gain the most advanced ideas found anywhere in the world. To the degree we cripple multinationals in the United States, we cut ourselves off from new ideas that will rapidly be shared among our competitors.

The *strategic* development of our policy is the key to our future. There are times when we *must* limit access to United States markets to force other countries to open up their systems. The Japanese have been the worst of all our trading partners, using a variety of laws, rules, and customs to hide their economy from foreign competition. This gives their domestic industries unfair advantages in international competition. Similarly, the European Common Market has recently initiated a whole series of protectionist moves to limit imports while at the same time dumping huge quantities of its agricultural surplus overseas at prices and with financing clearly designed to sell it below the cost of production. In such cases we have to be willing to retaliate in order to avoid being preyed upon.

I am not an advocate of the kind of blind, simple-minded adherence to free trade which crushed Great Britain between 1880 and 1914. Germany and the United States protected their industries behind high tariff walls and devoured British industries which were left open to exploitation by the British government's ideological commitment to free trade at all cost. A proper goal for a government—the British then and ours today—is the freest trade possible, given the behavior of your competitors.

But while tariffs are a reasonable, realistic tool of fair trade, a policy of protectionism would be ruinous. The drift toward protectionism is increased by every pandering Democratic speech which reassures workers that they do not have to learn new techniques, adopt new work rules, or bring in new labor-replacing technology such as robots.

One of the tragedies of the last twenty years has been the fact that the Liberalism which gives the most speeches about jobs has done more than any other element of our society to kill jobs. The mindset of the Left has combined work restriction, an anti-capital investment tax code, and a hatred of individual initiative with a hatred of high technology until there is no longer any basis on the Left for a rational program of modernizing America by using high technology.

C. P. Snow's concept of the Two Cultures, one scientific and the other based on the humanities, has come home with a vengeance. The engineers and managers who understand the objective requirements of a high technology future find themselves unable to articulate in moving, human terms their vision of a future with a higher standard of living and better jobs for everyone. Meanwhile, the poets, writers, and artists who can reach out so powerfully through films like *The China Syndrome* and *Norma Rae* have no understanding of the world of the future.

A cultural bias toward a more pessimistic, more defensive view of the future runs very deep in the American Left. As a consequence, the Left has been consistently wrong in recent years about the likelihood of jobs being created by technology.

Every major technological breakthrough, from internal-combustion engines to computers, has ultimately created jobs. Furthermore, the new jobs have ultimately been better and more humanizing than those of the previous culture. For that reason, people have consistently moved from jobs of an earlier era to those of the later era: plowing behind a mule was less desirable than working in a textile mill; working in a textile mill was less desirable than driving a long distance truck. For two-hundred years, people throughout the Western world have moved toward higher technology and better paying, cleaner, more human jobs.

The Liberal community is unable to accept that reality of progress through change for reasons both self-serving and cultural. The special interest of the men who run American labor unions has become fixed on maintaining quotas of extant jobs rather than on managing a transition to better jobs in new industries. Because of the historical structure which modern American labor unions inherited, it is to their disadvantage to encourage change in jobs. Every decline of an old industry and rise of a new one means dislocation for American labor unions.

Thus, American labor unions resemble their British counterparts in vigorously resisting technological change. The worst example of the old union mentality in both countries has been that of the printers unions which have killed dozens of newspapers in the name of protecting jobs.

Continental European and Japanese labor unions have less of the craft-and-confrontation mentality of the Anglo-Saxon tradition. As a consequence, they have had an easier time managing the transition to new jobs. Despite their relative advantages, however, even the European labor unions are remarkably static in their views on work and jobs. In almost every case, labor unions prefer to keep the jobs of the past rather than risk inventing the jobs of the future.

It is unnatural to expect people and institutions to voluntarily commit suicide. In that sense, maybe American unions are acting in the only way they can. However, an effort must be made at the national level to shift the focus of the unions toward growth in new job areas and toward helping their unionized companies manage the transition so they can survive in a changing environment.

Despite occasional real efforts—notably by the Communications Workers of America and the United Auto Workers—to shift values, it is extremely difficult for the union movement to modify its attitude of protecting the past and fighting change. The very nature of union elections tends to build in a pork-barrel, special-interest focus. This allows the most protectionist attitudes to flourish, because that is the kind of rhetoric which has come to dominate union elections and union politics. A reform candidate who argued for fewer benefits and more change would almost certainly be defeated in most union locals. For thirty years there has been a tradition of *more*. That tradition is hard to end even when the great advantages we inherited from World War II have been used up.

The biases against change of the labor union leadership are reinforced by the rhetoric of the Democratic political leadership. These are the politicians with whom the union leaders most frequently meet and by whom they are influenced in their analyses of the political-economic world. These liberal Democrats are, in turn, influenced by their own history: since Williams Jennings Bryan, they have been the anti-business, anti-managerial party. It is their tradition to blame management for whatever is wrong, just as it the union leadership's to assume that the industrial workers are already doing everything that could be asked of them and, therefore, any and all problems in the American economy or with corporations must be the fault of management.

The Democratic capacity to lead its supporters toward change is further weakened by the style of Bryan's populism: if your supporters are engaged in a class war with the evil forces of Wall Street and the country club, then you can hardly propose that America will only become healthier if everyone sits down at the same table and plans a program of economic health through common retraining and reinvestment.

Significantly, it is the American labor union leadership and Liberals who feel most threatened by the development of the Japanese management systems in the Nissan plant in Tennessee and the Honda plant in Ohio. If the Japanese succeed in breaking down the management-labor barriers which have been an integral part of American industrial warfare for over half a century, they will have dealt a severe blow to the Left Wing of American politics. Just as the British Labor Party relies on class hatred and class jealously to hold together its union support, so the American Left relies on the management-labor jealously in the American heavy industries to hold together its blue-collar voters.

Democrats face a real dilemma in that their own short-term interests may require support for policies which slow the rate of technological change, yet allow them to protect their allies from the perils of unemployment or movement into new industries. This problem has been increased by the fact that many of the technologically advanced and rapidly growing service industries are among those least susceptible to unionization precisely because their ranks are made up of self-confident people who see themselves as being at the cutting edge of technology and capable of managing their own changes. Such workers need no protection from their management through a union.

The next generation of smaller and weaker unions John Naisbitt describes in *Megatrends* means trouble for Democrats. Ironically, as unions have begun to lose strength in the society at large, they have taken very specific steps to increase their muscle in the Democratic party itself. Thus, 1984 is the first year the AFL-CIO is actively backing a candidate long before the primaries, in order to maximize its muscle in nominating a Presidential candidate and in writing a national platform.

These practical, short-term pressures that keep Democrats from embracing economic and technological change are pressures that are compounded by cultural influences.

The Left in Britain and America has been anti-technology since the 1880s, when the rise of Ruskin and other cultural critics of modern industry set the stage for a romantic rejection of high technology.

With increasing persistence, the Left has focused on the negative side effects of technological change and ignored the positive impact. Thus, radiation means the horrors of Hiroshima without consideration of the enormous contributions of X-ray technology to modern diagnostic medicine.

Because the Left has rejected any hope of salvation through technological innovation, Liberals led the fight against building the space shuttle in the early 1970s. Their argument was that research and development money was wasted because it had no immediate impact on Earth, while more money for food stamps was good because it provided an immediate benefit. This school of thought, which demands immediate gratification because it holds that there is no future, has gained force in the last decade.

Using this logic, it would have been wrong for the government to study bread mold. That would sound silly in a newspaper headline and would be easy to ridicule. Only after the study of bread mold led to the development of penicillin did we have proof that it was worth doing. In case after case, a seemingly ridiculous study has led to a great breakthrough.

The problem with research and development, especially in high technology, is that it is impossible to know in advance if your basic research is a good idea or a waste of time. The only way to guarantee fundamental breakthroughs is to investigate every path—most of which will indeed turn out to be blind alleys. Liberals find it impossible to understand the need either to accept blind alleys in research or the great prospect of a major breakthrough that could change the world. This negative cultural view of technology leads to a news media that focuses very intensely on the negative.

The forces of special-interest unionism and special-interest industries are combining with the Liberal wing of the Democratic Party to avoid change, to hide from competition, and to delay technology. If they are successful, America's future becomes bleaker. In their pessimistic view, for example, people would have a harder time working at home. The labor unions fight very hard to keep in force Labor Department work rules designed in the Depression to stop sweat shop tactics in cottage industries. These same rules can be used to block the emergence of new and desirable opportunities for working at home.

Our pessimistic future grows worse if we assume that the Liberal special-interest coalition pours more and more money into decaying bureaucracies. Instead of modernizing and trimming the current archaic government structure, it is likely that a Democratic administration would listen to the various government workers unions and expand the work force's outmoded systems rather than requiring it to change and learn new, more powerful systems.

A left-wing American movement would increase the benefits of welfare while, through higher taxes, decreasing the benefits of work- and job-creating individual initiative. The result would be a double whammy: unemployment becomes more desirable while employment becomes less desirable; marginal workers would have to act against their own best interest if they stayed off welfare; would-be entrepreneurs would find it easier to live off the capital they had already accumulated; the pace of new technology, new jobs, and new competitiveness would be slowed. The cumulative effect is an America far less able to compete with foreign countries and to create new work opportunities for our children.

As working and job-creation became less desirable, the pressures to raise money for the government's growing welfare programs would increase, resulting in a vicious circle: the more taxes were raised, the more people would go on welfare or decide to live off their savings, narrowing the tax base; as a result, there would be more pressure on the people who stayed active, initiating a downward spiral with enormous pressures for money to pay for all the left-wing programs; the Internal Revenue Service would then be required to consider every possible angle to get more money from the taxpayer to run the government.

The cycle in which we were trapped during the late 1970s would reassert itself as more and more people sought to escape high taxes through legal loopholes and by cheating. The pressure of high taxation would lead to a hemorrhage of people out of the tax system, again increasing the burden on honest citizens who try to remain faithful to their government—and even they would devote more and more time to "structuring" their income so as to minimize taxes. In real terms, honesty and patriotism would be punished by high taxation, while dishonesty and criminal behavior would be rewarded by larger after-tax incomes. Social norms would thus be reversed with good behavior punished and bad behavior rewarded.

The oppressive tax system would still be insufficient to pay for all the special-interest deals and welfare-state programs the left-wing politicians had promised their allies. Deficits would prove even larger than those we have already experienced. The pressure on the money markets to finance the debt would lead to higher and higher interest rates. We would rapidly regain Jimmy Carter's post-Civil War record of 21.5 percent interest rates. The economy would begin to shudder and collapse under the weight of a government larger than the public was willing to support.

In order to finance all its efforts, the Democrats would resort, as frightened politicians routinely do throughout the world, to printing paper money and inflating the currency. For a brief time inflation works because people accept the devalued money. Then they begin to realize that the dollar's real value is declining. Americans who felt in early 1980 that their world was collapsing with skyrocketing interest rates, high gold prices, and galloping inflation will have another chance to experience economic decay if the Left uses inflation to finance a really massive government program.

I keep worthless money in my office in Washington to remind me of the dangers of hyperinflation. The bills are from Weimar, Germany. They were given me by Ted Hirsch, a former mayor of Carrollton, Georgia. He is a German-born Jew who fled the chaos of Germany just before the rise of Adolph Hitler. Ted, a hard-working man who had made a good living out of the textile industry, helped manage the physical plant for West Georgia College and won the respect of his new countrymen sufficiently to be elected mayor of Carrollton for several terms.

After I won election to the Congress with Ted's support, he called me over to his house one afternoon, pulled out some Weimar currency which he had had for fifty years, and asked me to keep it as a reminder of the foolishness of politicians who think they can violate the laws of economic reality without being punished. He told me stories of his childhood—of trying to survive in a country in which inflation was crushing the common man, the same inflation which was creating great opportunities for the rich and paving the way for Adolph Hitler and the Nazis.

I have never forgotten the intense passion Ted felt for the role which honest money plays in creating an honest society. If your society promises you that a life of hard work and savings will pay off in honest money, then you can plan to spend your time accordingly, build up a pension fund for retirement, try to earn the cash to buy a few worldly things, and measure your progress against others in your generation by using a commonly accepted yardstick.

However, if money begins to decay in value, then the economic world is without a fixed standard of measurement. How can I know if I am making progress or losing ground if I have no reliable measure of value over time?

Imagine that we had a Federal Reserve Board for Distance and that each Friday it announced the length of the yard. One week a yard would be thirty-six inches; another week the yard would be thirty-two inches: occasionally, the yard would gain some of its former length, but the long-term trend would be for the yardstick to keep declining. Thus, since 1967, the yardstick would have shrunk to one-third of its former value. Today's yard would only be twelve inches long. I myself would today be eighteen feet tall.

While this analogy is intended to be funny, it is also a precise account of the impact of inflation on a standard of value called money. The average person is the person hit hardest by inflation: average people lack the resources to protect themselves against the sudden changes brought about by government. Historically, the very wealthy do fine in times of inflation and deflation. The common people are the ones who are crushed.

We should recall that it was a similar threat of inflation, brought on by Shays' Rebellion in 1786, which moved our Founding Fathers to create a stronger central government. Collectively they feared inflation so much that they were willing to go to considerable lengths to create a strong central government at the Federal level which would protect honest money.

The temptation to avoid saying no to some people by setting priorities runs very deep in human nature. As Lord Acton warned, power tends to corrupt. One element of that corruption is the willingness of powerful government and political figures to cheat their countrymen by printing money worth less in real value in order to avoid having to say no for the short run. In the long run, such a policy always has the potential for ruining a country. Gresham's law—that bad money drives out good—still holds true. Citizens learn that it is profitable to amass paper money and cheat others by buying real goods and services with phony currency. As productive people find themselves cheated by being given legal but phony money for real goods and services, they become less productive and less willing to exchange their hard work for someone else's paper.

Inflation almost invariably accelerates because people come to expect money to lose value and therefore begin discounting its value. The cancer grows with frightening speed. In our own history, the term "not worth a continental" was a derisory comment about the paper money printed by the Continental Congress under the Articles of Confederation. One of the first steps toward establishing a stable country was Alexander Hamilton's determination to redeem the honor of the United States by redeeming its currency. On that stable monetary system, a real boom in industry and trade developed.

There is a direct relationship between honest money and a serious crackdown on crime. Both these actions require a sense of absolutes, a sense that the world *has* to be run on the basis of some set of rules which transcend daily human desires.

A central characteristic of the welfare state is its refusal to subordinate any human desire to either organic rules of nature or to absolute rules of behavior. The welfare state is based on the idea that men are now smart enough to break economic, moral, and social laws with impunity. Gratification has become an adequate justification for any deviancy.

Thus, the effort to gratify all the special interests and ideological allies of any majority government would lead to pressure to inflate the currency in order to pay for more government, for more welfare, and for every other transfer program than the tax system could possibly accommodate.

While denying precisely those management and economic policies which would accommodate new technology and create more wealth, the welfare state insists upon spending more wealth than is currently available. The result is a short-term binge of government programs which, in the long run, leave the society actually poorer than it would have been with reasonable investment policies. This crippled the country in both the Lyndon Johnson and Jimmy Carter administrations. When taken to an extreme, it can actually bankrupt a society, as happened with Germany in the 1920s and with some Third World countries in the 1970s.

Faced with a massive government welfare program, steadily rising taxes, an ever more powerful Internal Revenue Service, and skyrocketing inflation, productive working Americans will regard the government as their enemy.

The greatest single change in American life, from Franklin Roosevelt's New Deal to the modern era, was the shift from government as a primary benefactor to government as a primary threat to the average person.

Among working Americans, the Internal Revenue Service and government bureaucracies in general are threats to everyday life. There are no Federal programs or agencies which are seen as commensurately helpful.

Modern populism has shifted from an anti-big business bias to an increasingly anti-government bureaucracy bias. One of the problems of modern left-wing reformers is that their solutions—more Federal agencies for one crisis or another—are falling on the ears of working Americans who are increasingly inclined to reject the Federal bureaucracy as an ally.

The populism of Proposition 13 in California and Proposition 2½ in Massachusetts has sprung up also in Michigan, where public reaction against a state tax increase by a Democratic governor and Democratic-controlled state legislature led to a recall petition and election in two suburban state Senate seats that resulted in two Republicans winning over 70 percent of the vote in largely blue-collar districts.

The populist reaction against high taxes was explained by one California policeman who favored Proposition 13, even though he thought it might cost him his job as local government cut back on local services: he reasoned that skyrocketing property taxes were certainly going to cost him the house he and his wife had dreamed about and worked for years to buy, but reasoned that he could get a new job more easily than he could buy a new house.

This tension between government and governed is brought about by three factors: first, even without inflation, a large welfare state inevitably intrudes on the pocketbooks of working Americans; second, the values of the Liberal governing elite are clearly different from the values of the working Americans they seek to govern—people are less and less willing to pay for the government they are getting; third, the very structure and nature of bureaucracy in its now-obsolete model make it so dysfunctional and disruptive to people in everyday life that they come to resent the very style and pattern of modern bureaucratic behavior.

These powerful wedges between the people and their government are now splitting the governed from the government so completely that they need to be examined more thoroughly.

First, whether the government takes money directly out of your pocket by raising taxes and indirectly out of your pocket by borrowing money and raising interest rates or devalues your wallet by adopting inflationary policies, big government is inevitably a big burden. Therefore, big government has to be remarkably efficient and responsive if it is to retain public support. Those government programs that focus on providing services, especially health care and aid for senior citizens, keep a reasonable level of public support as long as they are considered well managed. Countries like Germany and Japan, which have a tradition of a very professional civil service, tend to have reasonable public support for welfare also—although, in Japan, the intrusion of the government is also kept fairly low in financial impact.

The American welfare state has been built, in large part, around battle cries of helping the poor and reshaping society. It is therefore seen as more than simply a self-insurance program for working Americans: while people will pay a fair amount for their own health care and their own retirement (or that of their parents, on the assumption that their own children will continue to uphold and eventually benefit from the tradition), they will not pay nearly so cheerfully to sustain others. In times of slow economic growth, when their take-home pay is not increasing, workers will tend to become very angry if they feel they are working both to sustain their family and to sustain other people who are not carrying their fair share.

Thus, the Establishment Left in America has ambushed itself in its efforts to deal with working Americans. Liberal battle cries which are designed to rally the poor and ethnic minorities to support the Democratic Party by promising them government largesse have been translated by working Americans into battle cries against their own self-interest. Thus, every time Liberal candidates go on television explaining how their election will help specific voting blocs and specific special-interest groups, they alienate and arouse the anger of working taxpayers who have come to realize that government gives away only what it has first taken from the paychecks of workers.

This is a radical departure from the reasoning of Franklin Roosevelt in an age when both the income tax and the Social Security tax were very small and hardly felt by workers. Since then, inflation has pushed workers into higher brackets and the pressure of government spending has dramatically increased the load of both the income tax and the Social Security tax. As Jack Kemp has argued, the combination has changed the face of American politics.

By looking at the average American's tax burden under Harry Truman, the real scale of change can be appreciated. When Truman was President, the deduction for each member of a family was $500. Today, that deduction has only reached $1,000 despite all the inflation in the intervening years. If we were to provide the same deduction for dependents as Americans received after World War II, the deduction would have to be $4,800. It is no wonder that we see more divorces and more destructive pressure on the family when even the tax code has been biased against the survival of the traditional family unit.

The problem of Federal income tax is increased by the regressive nature of the Social Security tax. The Social Security tax hits the average worker three times:

First, it takes money directly out of their payroll before they ever see it. Today, for over half of all American workers, Social Security takes a larger bite out of their income than does the Federal income tax. In years to come, that bite will grow worse as the maximum tax keeps going up until, in 1995, you will pay 7.65 percent on $71,800 income. At that time, for most workers, the Social Security tax will be a larger burden than the Federal income tax.

Second, the Social Security tax hits workers hard because it takes money away from their companies that they never even see: every dollar a business has to pay in matching Social Security tax is a dollar it could have paid in additional income to the worker; in effect, the Social Security matching tax is a tax on hiring people.

Ironically, the Liberals who talk most about creating jobs have instead created a tax system which discourages business from hiring people because it raises the cost of hiring workers by 7 percent. Thus for every fourteen workers hired, the matching tax forces the business to give up hiring an additional person because their money has gone to paying the Federal government.

Furthermore, the matching tax is a severe burden for American companies competing in the international market. In effect, the Social Security tax is now a more than 6 percent burden on most American companies trying to compete with foreign manufacturers. Since Toyota does not have to pay American Social Security tax on the cars it builds, Toyota automatically has a significant price advantage over American companies. The average American automobile has 336 man hours involved in manufacturing it. Since those workers are paying the maximum Social Security tax and their company is matching it, each American-built car represents 213 dollars in Social Security tax alone.

The American competitive disadvantage in the international market due to the income and Social Security taxes is compounded by the fact that most of our international competitors have adopted a Value Added Tax. Value Added Taxes have the dual advantages of being taxes on consumption and of being refundable for exports under the GATT (General Agreement on Trade and Tariffs). Toyota gets a refund of between 17.5 percent and 22.5 percent on every car it sends overseas. Similarly, Mercedes gets a 13 percent refund on every car it ships outside Germany. The German and Japanese governments are thereby encouraging their citizens to export goods and to create jobs at home by selling products overseas. By contrast, the American tax system is not refundable under the GATT. American manufacturers are at the dual disadvantage in that their competitors get tax refunds while they themselves have to pay taxes. Thus, the very tax structure of the welfare state helps kill jobs in America while encouraging jobs overseas.

Finally, the Social Security tax hurts American workers because money taken out of the paycheck before it gets home limits the amount of income a worker can save. By their nature, consumption taxes encourage savings: if you save, you don't pay the tax. Income taxes and the American Social Security tax discourage savings: since you have already paid your taxes, there is no tax incentive to save.

Furthermore, since your paycheck shows an enormous government deduction for your Social Security tax, you have every incentive psychologically to assume that it is the government's job to take care of your retirement years. Therefore—both in practical take-home pay terms and in terms of a forthcoming retirement which you believe the government owes you—there is every reason for workers to believe the government has a contract to take care of them when they retire.

Thus, the welfare state builds the momentum of its own problems. Having driven a wedge into the worker's income and having discouraged exports and new jobs by its policies, it also encourages workers to depend on the government for their retirement funds. When workers do retire, they put maximum pressure on Social Security to keep raising the monthly payments. After all, having paid that money out of their paychecks and having looked at that deduction each week for years, workers come to believe that the government really owes them a rather large Social Security check.

In the welfare state, you can take a tax deduction if you borrow money, but you are taxed if you save money. This objective reality is a further bias against savings. Our government finds that its retirees have saved far less than they would have if they lived in Japan or Germany.

Japanese workers have been saving an average of 19 percent of their income over the last twenty years, while Americans have been saving an average of only 6 percent. Retired Americans must depend that much more on their government for pensions.

Since the pension pressures will be enormous and the government will not be able to tax a weakened economy (in part because the tax base is crumbling because of international competition), the government will try to borrow money to meet its obligations. However, the same spending habits which lead older Americans to rely disproportionately on their government also mean that there is not enough money to borrow.

The savings habits which better prepare Germans and Japanese for retirement give industries in those countries larger pools of capital from which to build new factories and create new jobs. These habits also have a tremendous impact on financing government debt: since the Japanese save three times as much as the Americans, they can lend their government much more money at a given interest rate without affecting the borrowing requirements of private companies.

By contrast, a limited savings rate harms Americans threefold: it does not provide enough savings for retirement and so increases the pressure on government pensions; it does not provide enough money for building new factories, developing new products, and creating new jobs; it does not provide enough liquidity to finance the government borrowing which is imposed by a weak economy and by the welfare state demands of pensioners and others.

Thus, the bad habits and bad structure of the welfare state come full circle and cause citizens to increasingly regard their own government as their adversary.

Most Americans have a tradition of saving and believe that they should save more than they do. We are not a country of big spenders; if anything, working Americans feel guilty about not saving enough and not preparing for their retirement years. Given the proper encouragement, Americans might well reverse the spending habits of the last thirty years and rival the Japanese as savers.

It was this underlying feeling about the propriety of saving which led to the outpouring of public feeling about the proposed bill to levy withholding on savings and interest, one of the greatest populist outcries of recent times in this country.

People reacted with remarkable ferocity to the proposal to withhold taxes on their interst payments. Although the view of the Washington elite on this issue was that it was all a banking-industry ploy to avoid paperwork, the reality I experienced as a representative holding town hall meetings in Georgia was remarkably different.

True, bankers did not want to get involved in withholding interest on savings, but retired citizens who had saved all their lives were livid. This was not just a rebellion of the very rich or of tax cheaters who had avoided paying their fair share; this was an uprising of everyday citizens who, at some fundamental level, had concluded that the government was going too far.

People felt personally offended. In part, this was because they really believed in the principle of saving and thought this was one more step to gouge the fruitful and the self-disciplined in order to spend money on the fraudulent welfare recipient. There was a sense that the government was once again punishing people who were doing the right thing while helping people who were doing the wrong thing.

In addition, people were enraged that an already too powerful and omnipresent Internal Revenue Service was now reaching into their savings accounts for every last dime. As a former history teacher, I kept reflecting on the attitude of the everyday farmers toward the French tax collectors during the last days of the French monarchy. The same sense of "us and them" came up in town hall meeting after town hall meeting which I attended. People were simply tired of being pushed and gouged for more money by their government.

Finally, people did not understand the complicated proposals by which those who were poor enough or had a small enough amount saved could sign a form and avoid the withholding tax. Some of those who *did* understand were outraged at the thought of having to tell their banker they were too poor to be taxed. These people saw it as a shameful badge only one step removed from going on welfare. All their lives they had worked to save enough to have their banker respect them. Now they were being told by their government they should go to their banker and plead poverty to avoid being taxed. Those who did not understand the procedures assumed that, when in doubt, it was better to distrust the government.

It is the latter attitude which so reveals the gap

between the governing elite and the conservative nation which is struggling to rise and seize power. No one I talked to on the Ways and Means Committee or in the Treasury understood the intensity of distrust which now exists in the country at large. Sincere, well-meaning men kept explaining that they were protecting the small saver and that there was no real ground for complaint. They could not understand that the country now distrusts the government's tax collectors so much that, when in doubt, it disbelieves its own government.

The rebellion against withholding on savings interest is a suggestion of the working American rebellion which is brewing throughout the country. If the welfare state is re-established, its policies of higher taxation, a badly structured tax code, and inflation will lead to a revolt on a scale we have not seen in this country in the 20th century.

But the real danger is that the Establishment Left will be neither replaced nor reformed. If Liberalism regains its hold on Washington and is able to pass laws taxing the public, it nonetheless will find it extremely difficult to coerce a nation which does not believe in the ideological purposes of the welfare state and which does not believe in the fairness of the tax code. Even a totalitarian state like the Soviet Union finds itself beset with so much corruption that it must maintain enormous police forces and resort to draconian punishment—including executions—for crimes like black marketeering.

A free society like ours relies heavily on the support of the public to enforce the law. That requires that the laws be considered fair and just, and that the public thinks of the government as "our government." The minute the public begins to think of the government as "them," the capacity of a free society to enforce its tax code and to maintain a lawful economy decays rapidly.

The rapid spread of corruption in the tax code will lead to a similar spread of corruption in other avenues of life. If you are accepting cash so you don't have to pay taxes, how can you turn in your friend for dealing in cocaine or for bribing the building inspector or for anything else short of violent crime? Once it becomes acceptable for us to break the law in an area which we agree is unfair and unjust, then we replace all social rules with personal judgments. Down that road lies a level of corruption which is dangerous for a free society.

In recent years, we have had a number of cases in which law enforcement officials were corrupted by dealing with the high-profit cocaine and heroin traffic. A number of the law enforcement officials who went on the take in narcotics found it easy to accept levels of crime in other areas which they would never have condoned prior to their own corruption.

Corruption can spread until entire communities are governed by alliances between various illegal interests, and the innocent, law-abiding citizen must survive in obscurity by avoiding contact with the various powerful forces of corruption—of which the formal government is one. The combination of an unfair, increasingly burdensome tax code, the drug culture, and the ethics of the welfare state encourages the kind of decay which makes it plausible for honest citizens to become dishonest.

Even in areas where the welfare state has tried to be helpful, it finds that general corruption seeps in to undermine its efforts. There are serious problems with food stamp fraud, with a black market in food stamps, and with counterfeit food stamps. Whenever there is a commodity that can be used as currency, there is a temptation to counterfeit it. Thus the welfare state has to cope with both counterfeiters of real money and counterfeiters of welfare money.

When people cheat on their taxes, on food stamps, and on drug laws, it is little wonder that we see examples of cheating in government contracts. This is true at the level of the defense industry when some major contractor defrauds the Pentagon, but is also true locally when some unscrupulous doctor overcharges for Medicare and Medicaid treatments. Periodically, we are treated to the spectacle of the news media or a Congressional committee uncovering some company or individual who has stolen money from the people of the United States by over-charging or cheating on a government contract. Routinely, the call is raised for more red tape, more bureaucracy, more regulation to ensure that such cheating of the taxpayers can't happen again.

The result of seeking honesty through bureaucratic enforcement is that the honest contractor or doctor becomes burdened with even more paperwork; anger against the bureaucracy increases; the truly wicked find new and even more complicated systems for cheating since loopholes are inevitable in any human system; the burden of enforcement grows heavier, and the likelihood of actually prosecuting any single case grows less as the weight of enforcement becomes overwhelming.

We are now trapped in a downward spiral as the structure of welfare-state bureaucracy alienates citizens and makes them less likely to obey the law: the more we rely on red tape, the weaker grows self-enforcement; the more we rely on outside enforcement by bureaucrats, the less citizens believe that they have an obligation to discipline themselves. Just as most Americans exceed the speed limit when they are driving, so the proliferation of red tape and bureaucracy leads to a proliferation of cheating. Ultimately, the burden will become too great and our society will resemble the decaying Ottoman Empire rather than the honest, open, system we have known as the American Way. Such is the scale of threat which exists in the current decay of our value system and the decay of our public morality.

The welfare state's structure not only encourages illegal acts among some citizens, it also encourages a sense of powerlessness in virtually every citizen. The structure of a large centralized bureaucracy guarantees that most people will come to feel that their time and efforts are wasted. Most of us cannot cope with giant systems and big structures. Furthermore, most of our concerns don't justify the effort that is required to modify a large centralized bureaucracy.

People have specific problems. Whether they involve their children in school, or their parents in a nursing home, or a specific government regulation that may relate to opening up a business or building a house, they want to be able to get the job done. Just as people can become excited about actually solving a problem or finishing a job, they become exceedingly frustrated when they are faced with a maze of red tape, an enormously complicated system of rules and procedures, or a very large structure without clear lines of authority.

Public education, in particular, has suffered a loss of public interest and support because of the complex system of local boards of education, professionalism through accreditation systems and schools of education, and Federal and state bureaucracies—a combination which guarantees that whatever questions parents ask will not be answered.

It is exceedingly difficult to change a textbook or a promotion policy, to get a bad teacher fired or a new course adopted, or to protest the way in which a particular topic is taught. Any parent who has tried to work with a large public school system knows that the answer to virtually every question is, "My hands are tied."

Accreditation belongs to a private agency. The system of teaching is the province of the state college of education. The special funding for specific programs comes from the Federal government with strings attached. Parents are understandably frustrated and feel powerless. Powerlessness leads to withdrawal and apathy. Withdrawal and apathy lead to alienation. Suddenly, we once again have an "us versus them" rift building in a free society in which "them" should *be* us.

I have personally experienced the frustrations as a parent, a part-time teacher in a public high school, and a professor in a state college. I have talked with active school board members who, after years of effort, have come to feel impotent and who believe that the real answers lie in a state legislature. When even school board members can feel powerless, it is a sign that the structure of the bureaucracy has become unwieldy, and perhaps even destructive, to the very habits of good citizenship in a free society.

It is vital that we recognize how debilitating centralized command bureaucracies can be to a free society. This "centralized command bureaucracy" model was developed during World War I to allow countries to fight a life-and-death struggle. When used for a few years, in a time of national crisis, with all the drive and spirit of patriotism behind it, a command bureaucracy can be an awesome instrument for the organization and direction of a free society.

Americans proved in World War II that free people can voluntarily subordinate themselves to a centralized government bureaucracy for the duration of a crisis and produce remarkable results. Many historians believe that the United States actually mobilized far more of its potential power than did Nazi Germany or Imperial Japan.

But the key to a free society's use of a command bureaucracy is that this use must be clearly understood as short-term, and clearly and intelligently directed toward achieving nationally agreed-upon goals.

Once a bureaucracy is allowed to exist beyond the emergency, it rapidly develops its own agenda and its own internal politics. The spirit of patriotism wanes as the spirits of organizational territoriality and self-interest gain strength. Presently, the general public begins to realize that demands which once were made in the national interest are increasingly being made in the narrow, selfish interest of the bureaucrats. The loyal spirit of cooperation is replaced with a sullen hostility which looks askance at all requests of the bureaucracy. Finally, the citizenry grows cynical and decides to adopt whatever mechanisms are necessary to avoid effective surveillance by the bureaucracy. The bureaucracy, confusing its interests with the national interest, responds by adopting increasingly rigorous rules, regulations, and red tape to enforce its will on an unwilling populace.

While this process has most clearly occurred with the Internal Revenue Service, it is important to recognize that virtually every area of Federal, state, and local bureaucracies have now the same patterns of red tape, leading to surface compliance by citizens who scheme and plot to do as they please without regard to laws made by an over-regulatory bureaucracy.

The underground economy of the tax system is matched by underground behaviors in public housing, in health care, in education, in welfare, and in virtually every phase of American life. The independent spirit of America has not been tamed, but merely tempered by a cunning pretense of obedience while actually working to get its own way. While this underground response to bureaucratic rules has allowed many citizens to prosper despite Washington, it has not provided an answer to the general problem of changing or reforming large bureaucracies.

One of the most worrisome symptoms of this problem is the degree to which large bureaucracies have brought people to the point of apathy or even alienation. People believe you can hide from the system a large part of the time, but they don't think you can really change the system. One of the reasons Americans don't go to school board meetings or hospital meetings, or even vote, is that they believe the process, now so complicated and unwieldy, makes their effort irrelevant.

The result of years of large structures running our government at the Federal, state, and local levels has been an increasing underground activity coupled with an increasing withdrawal and apathy.

Liberalism continues to seek answers through the imposition of large public bureaucracies. The allies of left-wing Democrats are largely the unions and associations which support centralized bureaucracies. It is therefore impossible for Democrats to propose reforms which would allow people to govern themselves locally again, or to propose changes which would make public services sufficiently responsive to the citizen-as-customer reforms so that people would again become excited and involved in their local education, health, and other government functions.

Short of building a new model of citizen-as-customer-oriented, decentralized government such as the opportunity society proposes, I don't see any solution to the alientation and anger building in the voters of this country.

When you watch Democrats run for President there is no sign that they or their major allies appreciate the breakdown in public confidence in government. Therefore they are demanding more, rather than less, of the structures and habits which most Americans now reject.

It is conceivable that the Democratic Party will someday find a new, acceptable solution to the issues of American life; but it will do so only by repudiating the protectionist, technophobic, centralized model which currently dominates its activists and allies.

Let me bring together, in summary, the negative synergism of the welfare state as it multiplies the defects of bad tax policy by bad economic policy and bad social policy.

In an effort to defend the bureaucracies, industries, and special interests of the past, the welfare-state politicians and bureaucrats adopt policies which make it more difficult to develop the vision, energy and commitment which are necessary to transform America from an industrial society into an information society.

By resisting virtually all of the trends John Naisbitt suggests are inevitable in *Megatrends*, the welfare state ensures that our industries will not be able to compete with foreign industries in the future. Then, in order to prop up our decaying industries, welfare-state politicians will raise tariffs and quotas which will lead to trade wars, which, in turn, will weaken the Western Alliance and increase the opportunities for Soviet adventurism.

As our industries decay, costs for the consumer go up behind a trade barrier; the burden on the welfare state grows greater; higher taxes increase tax cheating. This combination of pressures leads to an increasingly corrupt society in which people break the laws they dislike.

Faced with a declining revenue base as people get better at cheating, welfare-state politicians will be reduced to inflating the currency in order to feed their various allies. This leaves American industries even less competitive with overseas companies, while raising the size of the budget sharply to cope with cost-of-living allowances for pensions, welfare, and other programs which drive up government spending. As American workers try to maintain their standard of living, they fight for inflationary wage increases which further endanger their jobs. This, again, increases the size of the welfare-state budget and the pressure to keep inflating the currency.

Confronted by demands for more and more from their welfare-state allies, Liberal politicians try to control skyrocketing deficits by cutting their investment in defense, in transportation, and in other infrastructures. There are no Liberal allies in these two areas, so Liberals are able to cut both these categories without suffering much political pain. Both cuts weaken the nation's future.

Further, the welfare-state politicians find themselves unable to offer serious reforms for any of the major systems of government activity. Trapped by their own allies into supporting the existing bureaucracy or changing it only by adding more personnel and raising salaries, the welfare-state politicians are unable to seriously challenge the extant status quo in education, welfare, health, or the law. Whenever a serious proposal for change emerges, the threatened interest groups reach the welfare-state politicians they support and block the reforms. It is permissible to de-regulate non-welfare state functions such as trucking and airlines, but any effort to overhaul the big areas of public life is anathema to the interest groups involved. They are thus taboo for any Liberal politician to touch.

Since this downward spiral of inflation, protectionism, bureaucracy, and decay has no high-technology escape clause, there are no real grounds for hope in the welfare-state ideology. Liberal politicians rapidly become gloomy and pessimistic. When they assess the future, they find only more inflation, more decay, more unemployment, and more protectionism.

As the rest of the world watches a decaying America grow sicker economically, more pessimistic politically, and weaker militarily, the world becomes a more dangerous place. The forces of terrorism and violence come to believe they can take risks with impunity. The forces of freedom and stability feel their power diminished by the absence of an effective United States.

It is possible to look at this dismal projection and conclude that it is just the political rhetoric of a conservative Republican Congressman. However, I have a simple challenge for the reader:

I say that every single negative event outlined in the summary on the welfare state actually occurred under Jimmy Carter. Every single problem had appeared by the end of 1980. The country rejected Carter's bid for re-election in part because there was a feeling that the ship of state was foundering. The country was right. Left-wing statist policies not only did not work; they were a disaster of the first order.

Democrats have made the excuse that the crisis of the Carter years was caused by the personality defects of Jimmy Carter and his personal appointees, ignoring the fact that the Democrats controlled both the House and the Senate during those four years. The Democratic Cabinet Carter appointed included men trained by Lyndon Johnson and schooled in the Democratic Congresses.

Carter did not preside over an aberration of national Democratic policies; he embodied them faithfully. He did not lead a right-wing populist rebellion trying to cut taxes and dismantle the Federal bureaucracy; he may have campaigned as the most conservative Democratic Presidential nominee since Grover Cleveland but his answers to energy, to education—to social problems in general—were to call again and again on government in Washington. His plan to control health care costs was a classic example.

The failures of 1977 to 1980 were the failures of the big-spending, high-tax, high-inflation, centralized bureaucracy which I refer to as the welfare state.

There is no reason to believe that another Democratic administration could do better than Carter's did. Considering the promises of the Democratic campaign to its various allies, it is very likely that the next Liberal administration will be an even bigger disaster, even more dedicated to policies that have already failed.

The frightening thing is that our next such disaster could be our last: America is running out of time. After World War II we had enormous margins for error in our economic, social, foreign, and military policies. We were so powerful and the world so disarrayed that we could afford grand experiments here at home.

Now we face a more dangerous world with a razor-thin margin for error. I sat on the floor of the House of Representatives in January, 1980, and heard President Carter imply in his State of the Union message that the United States would use theater nuclear weapons to stop the Soviet Union if it followed up its Afghanistan invasion with an attack on the oil resources of the Persian Gulf. That genuinely frightened me. We had been reduced in real military power to such a degree that Carter had to verge on hysteria to communicate to the Soviets that we would protect the free world's oil supply.

We can no longer afford left-wing experiments in mismanaging the economy through inept policies and mismanaging the country through overgrown bureaucracy: the world is too dangerous. Our survival requires that we confront honestly the dangers of the world in which we now live. We must put our domestic house in order so that freedom can survive in peace through our strength.

Our domestic welfare policies have played around with flimsy houses of straw and wood because the Liberal elite really did not believe there were any wolves in the world. Since the world was benign, we could afford to focus our resources on the redistribution of existing wealth instead of the creation of new wealth.

It is time now to build a solid house. We must roll up our sleeves and rebuild an economically strong, high-technology oriented America capable of surviving any potential wolves. There is a tomorrow, and we will have to live in it. Those who enjoy life without realistic precautions may not enjoy life long.

For the last few decades we have been mired in ineffective welfare-state policies. If we want to lead the world to security and freedom, we must first get our own house in order. A strong foundation requires the building of an opportunity society based on fundamental principles that are rooted deeply in American history. It is to that task that we must now turn.

CHAPTER 5: PRINCIPLES OF AN OPPORTUNITY SOCIETY

There are some ideas so complex and so important that they both require and repay time spent in developing and understanding them. It is not enough simply to gaze through the window of opportunity at a desirable future, or into the shadowy world of the welfare state at a bleak future, and decide that progress is preferable to stagnation, that happiness is preferable to unhappiness. We must find a way to grasp what now we can only glimpse, to make our vision a reality. To do this, we must take the time to learn some fundamentals before we make basic decisions about the world we will give our children and grandchildren.

I have described and outlined the characteristics of the welfare state, the dangers of an increasingly violent world which demand that a dynamic America lead the human race into a future of freedom and peace, the exciting technologies which can make an America of hope and opportunity possible.

The transition from decay and violence to prosperity and peace requires a new approach to the structure of government and a reorganization of national activities—a fundamental "paradigm shift" from the defeatist, welfare-state mentality to one of focused determination to create an opportunity society.

Thomas Kuhn, in *The Structure of Scientific Revolutions,* developed the concept of paradigm shift, asserting that, when an existing system of thought fails to explain new observations about nature, then pioneering young scientists invent a new framework. It is this leap from an outmoded, and therefore weak, model of thought to an innovative, powerful model of thought which we later identify as a scientific revolution.

In Kuhn's thesis, the shift from the physics of Sir Isaac Newton to the physics of Albert Einstein is a typical paradigm shift. As physics and mathematics made it possible to frame questions which Newtonian physics could not encompass, it was necessary to create a new way of thinking about the universe.

The time has come to us to build a new model of American government that is more potent than the welfare state. The key to this approach is a paradigm shift to a more appropriate framework of thought rather than further modification of the outmoded and ineffective welfare-state model.

The basic changes I am proposing are simple:

from liberal	to conservative
from welfare	to opportunity
from the state	to the society at large

Consider first the values of Liberalism as a cultural force rather than a political phenomenon. Any anthropologist could visit various societies and describe the underlying values as "liberal" or "conservative."

If the students in a school sit neatly in rows and take notes obediently as the teacher talks, we tend to consider that a conservative society. If discipline is loose, if the students are encouraged to experiment on their own, and if there is a sense of openness rather than discipline, we would consider that a liberal society.

In my opinion, the right direction for America's 21st century will involve a shift back toward traditional values and conservative basic principles such as work, thrift, and a strong criminal justice system.

An opportunity society must, by definition, shift the emphasis from welfare to opportunities. The welfare state insists on seeing people as victims and government as the provider of help for their plight. A good welfare-state politician can walk into a room and identify every person there as a potential victim.

Consider these examples:

A dairy farmer whose industry is now so productive that, if not protected, he will be forced out of business is viewed by the welfare state as a victim of the very success of modern agriculture. His protector, the welfare state, levies taxes on the general public to subsidize his business, in the process restricting the amount of milk, cheese, and butter produced and thus artificially raising the price of dairy goods in the grocery store.

The next person is a worker who, taxed to the hilt by the welfare state, finds it difficult to survive on his weekly take-home pay and, because of the welfare-state dairy program, has trouble affording enough milk for his children. To this "victim," the welfare state promptly offers food stamps to cure a problem that it—not the worker/victim—has created.

Next we have a young couple, working to save enough money to buy a house, who find that—after paying the taxes necessary for the dairy program, food stamps, and the high price of groceries established by the welfare-state policy—they cannot save enough money to buy a house. The welfare state has a solution available to its such victims: with money taken from the worker and the dairy farmer, it has established a government loan program to put the young couple into a new home.

Each of the aforementioned examples cross-subsidizes the others, and all vote to keep welfare-state politicians in power.

The problem with this focus on welfare is that it forces people to spend their creative energies on manipulating the government rather than on substantive solutions. The welfare mentality encourages us to see ourselves as weak and helpless, potential victims unless we find a way to get the government to help us and protect us.

The politics of welfare and of victimization rapidly degenerate into the politics of fear. Virtually half of the Democratic Party's 1982 campaign was designed to scare senior citizens into believing that they were being mugged by the Reagan Administration in particular and the Republican Party in general.

While the Democrats controlled the White House,

the Senate, and the House in 1977, they claimed to have solved Social Security problems until the 21st century. I was still teaching at West Georgia College at the time; neither I nor my party had done anything to create the crisis in Social Security or to pass any of the laws that had led to an enormous shortfall in the Social Security fund. Therefore, I was amazed, an 1982, to be attacked for trying to debate and establish a fair system which both protects our grandparents and avoids crushing our grandchildren, a solution which would ensure that none of our senior citizens were hurt—but also that taxes would not be raised to such a level that our descendants could not buy new homes or save money for retirement.

In the midst of the fight over Social Security, I understood for the first time the irony of welfare-state rhetoric. Where does the image of poor, helpless, victimized grandparents come from? Here were Claude Pepper, at eighty-three, the new Chairman of the Rules Committee; Ronald Reagan, at seventy-two, President of the United States; and Tip O'Neill, at seventy, Speaker of the House—all fighting to protect helpless senior citizens. I, aged forty, reflected that these were three of the most active and powerful men in America—far more powerful than me.

An opportunity society would reverse the welfare state's focus on victimization. Most Americans instinctively believe in the dictum Peter Drucker has enunciated: that real progress is made by building from strength. Drucker argues that all great managers know that focusing on weakness is a sure road to ruin because weaknesses expand to absorb the time and energy devoted to them. Individuals who focus on their strengths liberate themselves and are then free to build lives of dignity.

A focus on strengths and opportunities would transmute our culture's current pathologically negative focus into optimism. One of the reasons *U.S.A. Today* has done so well as a new newspaper has been its upbeat, optimistic vision of the future.

Focusing on the strengths and assets of a community, an industry, or an individual raises the potential for positive change. We have all seen new coaches take over and change the fortune of a football team, or seen new managers turn a failing business around. In almost every case, these successes are stories of raising morale, painting a powerful vision of a better future, teaching positive habits, and emphasizing existing strengths which had been neglected previously.

Liberal politicians have difficulty focusing on the positive because their only stock in trade is "new" government goods and services. Because they see politics as a transaction between the voter and the government in which politicians get re-elected for having provided more goods to each voter, they continually seek new angles to "help" their constituency of perennial victims. In a very real sense, a Liberal politician would be lost in a healthy, happy community where all citizens had jobs and were buying their own homes without government help. The Liberal politician would have nothing to offer in a transaction for votes.

The opportunity society is almost the reverse of the welfare state: by focusing on opportunity and strength rather than victims and welfare, we create an expanding range of possibilities; each success creates the arena for future successes; by using the entire society as a framework for development, we open up options and opportunities unthinkable inside the welfare-state framework.

The welfare state is limited to the central bureaucracy as a resource. The opportunity society begins with each of us as individuals by exploring strengths in our family and neighborhood, calls upon volunteer associations (including churches), looks upon public agencies, examines the power and strength of the free enterprise market to develop solutions, and only then turns to government—first local, then state, and, as a final recourse, the Federal bureaucracy.

A free society is strong and durable because of its ability to solve problems and create opportunities at many levels. One of the great dangers of the welfare state is its degree of dependence on Federal bureaucracies. In an opportunity society, where the entire society is considered a resource, the resultant range of possibilities dwarfs those available within the bureaucratic state.

In a welfare state, the President is the head of the government and only occasionally talks to the people, normally and nearly exclusively about government matters.

In an opportunity society, the President leads citizens, themselves an integral part of the problem-solving process, sharing with them a vision of the future to which all subscribe and contribute to creating. Theodore Roosevelt's description of the Presidency as a "bully pulpit" was clearly aimed at an opportunity-society vision of America. Alexis de Toqueville's description of America focused on our strengths as a culture rather than the power of our government.

The great failure of Republican administrations in the past sixteen years has been their failure to understand that running a cheap welfare state cannot be our goal. We must commit ourselves to creating a new vision and a new paradigm of how Americans will live and work.

To shift a nation from one political framework to another demands a redefinition of fundamental principles. It requires first that political leaders, analysts, and citizens stop bandaging old problems and start finding new solutions. Franklin Delano Roosevelt catalyzed such a fundamental redirection when he created the New Deal by proposing that government play a vastly more active role in responding to the crisis of the Depression.

Today we need a new fundamental shift in political principles as we come to the end of the era of centralized welfare-state solutions. To understand the scale of change necessary, it is useful to look at the levels of activity which are necessarily involved in competitive behavior.

The hierarchical structure of human endeavor includes:

1. Vision.
2. Strategy.

3. Operations.
4. Tactics.

The great failure of the modern Republican Party, including much of the Reagan Administration, has been in its effort to solve problems operationally or tactically within the framework of the welfare state. Trying to run a welfare program more cheaply or trying to run a small-business loan program more efficiently will not change the world.

Real change occurs first at the levels of vision and strategy. It is important to realize that vision must precede strategy, strategy precede operations, and operations precede tactics.

While these terms originally came out of the military, they apply to the business and political worlds as accurately. Consider Peter Drucker's point that Henry Ford was never a small businessman, but from the outset a visionary who of necessity started with a "baby" business: the difference between small businesses and baby businesses is that the founders of baby businesses have large visions. McDonald's and Kentucky Fried Chicken are two examples—there were other hamburger stands and other fried chicken restaurants, but their founders lacked the vision of a Ray Kroc or Harlan Sanders.

Vision by itself, however, is not enough. As Drucker notes, plans have to evolve into processes, and that requires the development of specific strategies. The general concept of popular hamburger chains must acquire specific symbols such as the "Golden Arches" and specific quality control methods which apply to every McDonald's in the world.

Strategies cannot be managed by visionaries unless they can be planned as specific operations and turned over to managers who run decentralized operations. The most powerful model of this in American history was General George Marshall's delegation of enormous power to men like Dwight Eisenhower. Marshall kept strong control of the supreme strategy of global war-fighting, but he knew that individual leaders on the spot must have command of each project—even if the project was as big as the Normandy invasion.

At the very end of this hierarchy of activities comes the specific habits and behaviors which are expected from the army platoon sergeant, the airline flight attendant, or from the local McDonald's store manager. In the army, men's lives depend on their platoon sergeant and soldiers have a right to expect a certain degree of trained competence in their leaders—without it, trust and discipline breaks down. In civilian life, parallels abound in successful businesses: have you ever noticed that you are greeted almost identically in every McDonald's you visit or that you get a very similar description of aircraft safety devices on virtually every commercial airliner you ride?

The welfare state has a very specific vision of rational men and women working in a centralized bureaucracy to provide goods and services to victims who would otherwise not have food, housing, health care, or education—a model based on the great success of centralized bureaucracies in World War I. The efforts of negative conservatism to shrink the welfare state by cheapness or by fighting it at the tactical and operational level are doomed to failure: as long as the welfare state remains the dominant vision of our world, people will vote for a positive, generous Liberal party because, as a nation, we are neither negative nor cheap.

The failure of the welfare state requires that we change our national presumptions, our hopes, and dreams at the visionary level, then implement our new vision with new strategies and, finally, develop new operations and tactics for the implementation of those strategies.

The creation of a new pattern of problem-solving for a country our size is not the work of one man or even a hundred men and women. A free society remolds itself only through the concerted efforts of many people focused on a single goal.

There are a number of examples in American history which show how a free people can re-examine their assumptions about government and society. Thomas Jefferson argued that every generation needs its own revolution. We have been fortunate in that American revolutions have generally been fought at the ballot box rather than on the battlefield.

We have seen the development of Jacksonian Democracy in response to the frontier; the rise of Bryanism in response to agrarian economic dislocation; the McKinley-Theodore Roosevelt vision of full-dinner-pail, working America in response to the industrial revolution; the rise of LaFollete progressivism in an attempt to reform government early in this century; and the development of New Deal centralized government under F.D.R. The development of an opportunity society over the next decade will be a comparable effort yielding comparable, if not greater, rewards.

Just as earlier adaptations of American political thought were responses to changing times, so the opportunity society effort is a response to the revolutions in high technology. Just as earlier political movements were shaped by the values of a new generation of voters such as Jackson's frontiersmen and McKinley's industrial workers, the opportunity society will answer the cries of the baby-boom generation for a new politics responsive to the future's needs.

THE INTELLECTUAL PRINCIPLES BEHIND AN OPPORTUNITY SOCIETY

It is an affectation of the American Left that only *it* can have really new ideas. Yet its ideas reflect a bias against the passions which formed this nation and precipitated the American Revolution, during which our patriots were, almost to a man, anti-statist as well as pro-individual liberties and free markets.

The major ideas on which our opportunity society effort will be built all oppose the underlying concepts of the welfare state:

1. The world can only be understood by conceiving all its parts as an interactive whole.

2. Behavior must be considered as a system so that we can comprehend the second- and third-order effects of ideas, inventions, and structures.
3. We must realize that humans are organic beings when proposing plans and activities that affect people.
4. Effective political planning must be directed toward the future but rooted in the past.
5. Long-term thinking is usually preferable to short-term thinking.
6. No matter how powerful our nation is, it exists within an even larger world community.
7. There are certain canons of human behavior which have applied throughout recorded history.
8. A free society must be built on the strength and participation of its citizens.
9. A free society is founded on honesty, ethics, and the free exchange of ideas among its people and communities.

These principles are the core of an opportunity society. Taken together, they form a framework of analysis which is radically different from the welfare state's, creating a new stretegic base for problem-solving and the development of new opportunities.

Let us now take each principle by itself, explore its analogue in the welfare state, and see how each applies to a few examples.

1. *Holistic versus reductionist views of reality*

The greatest single weakness of the welfare state has been its infatuation with reductionism, the 20th-century scientific method of focusing narrowly on each part of a problem. Reductionism has worked very well in the sciences—there are specialists in the chemistry of the liver and specialists in specific aspects of subatomic particles without whose efforts we would know less about our world and how to survive in it.

If your goal is to build a compendium of knowledge about the parts of the universe, a reductionist approach is enormously powerful. When applied to management in World War I by governments intent on maximizing short-term production by breaking problems into their component parts and then solving each portion, reductionism yielded an enormous increase in the capacity to get things done.

The problem with reductionism is that it is essentially a short-term, narrow viewpoint which works only for specific goals at a specific point in time. The longer you apply reductionist methods in management and politics, the more out of touch with the entirety of the problem and the proportional value of its components you become. When Lt. Col. Dennis M. Drew, in "Military Art & the American Tradition," analyzed America's strengths and weaknesses as a military power, he concluded that, at the tactics and operations levels, we were enormously powerful: our reductionist style of management could deliver the goods for any specifically defined immediate goal. Drew further concluded that, at the vision and strategy levels, we tended to be ineffective because the reductionist management model is too limited and too short-sighted to be responsive to the requirements of the vision and strategy levels.

Col. Harry Sommers, in *On Strategy,* his study of Vietnam in the context of Clausewitz's *On War,* came to a similar conclusion. Sommers noted that MacNamara had brought the American management model to the Pentagon, a model brilliantly designed for war preparation, but disastrous in war *fighting*. Both studies agree that our cultural style, although adequate at the bottom levels of the hierarchy, crippled us at the higher levels.

Peters and Waterman, in the opening chapter of *In Search of Excellence,* similarly indict the American management system, arguing that the Harvard Business School system gains short-term power by coercing people in ways which build up such intense resistance over the long term that ultimately they destroy the business they were designed to manage.

The reductionist approaches of the welfare state lead to a whole series of absurd specializations which are clearly out of touch with reality.

For example, Health and Human Services deal with

children as centers of disease and as recipients of welfare; they are recreation users to the Interior Department; objects to be housed to the Department of Housing and Urban Development; nutrition consumers to the Department of Agriculture; sources of crime or its victims to the Justice Department; and subjects for education to the Department of Education. A real child is not a collection of six or more separate functions. A child is a whole being, and his or her needs can best be dealt with in entirety.

Furthermore, the welfare state's emphasis on reductionism eliminates the spiritual side of life because its addiction to a secular analysis of life is ultimately both degrading and destructive.

As whole beings in a holistic world, we need spiritual understanding and spiritual commitment. Faith can sustain people through enormous deprivation and pain, as the Jewish community has proven for three thousand years. Rich cocaine addicts prove every day that a secular youth with a great future but no beliefs can self-destruct in the midst of plenty.

By trying to solve the problems of poverty without considering the spiritual element, and by dividing human beings into component traits instead of regarding them as individuals, the various welfare-state bureaucracies have compounded the problems they hoped to solve.

Foreign aid is a classic example of the welfare state's myopia and its destructiveness toward people: the two most important exports we have are the Declaration of Independence and the free market system, yet, for two generations, the U.S. aid program has exported only money and government bureaucracy. Luckily, multinational corporations and student exchange programs have educated many Third World countries despite inappropriate U.S. aid projects, many of which the recipients could not maintain once we built them.

If you examine Liberalism section by section, its platforms and programs make some sense—it does offer more to each voter in one narrow zone or another. However, when you examine the totality of life, Liberalism is a remarkably destructive and empty political approach.

An opportunity society can look at the world, at society, at neighborhoods, and at individuals as wholes, recognize the spiritual aspects of life and the importance of dreams in motivating and directing people, then try to develop unified approaches that treat each of us as individuals, as families, as neighborhoods, and as communities.

In short, implementing the opportunity society would reorganize the government programs which should support and help individuals, families, and communities—rather than trying to reorganize people so they fit the programs.

Finally, the opportunity society would try to develop among intellectuals and academicians a concern for the totality of life rather than a too-fine focus on what narrow facets of life happen to fit their specialized studies.

The advice of reductionists has actually harmed Washington decision-makers and analysts by focusing their attention on narrow and misleading issues. We have allowed our attention to be misdirected by academicians who reshaped the data to fit their theories instead of reshaping their theories to fit reality.

Only the dominance of the reductionist model, combined with the mystical aura of modern mathematics, could explain the decay of political philosophy, which has shrunk from a 19th-century concern of substance into a late 20th-century form of esoteric debate among academicians. The computerized mathematical formulas of modern political philosophers relate less and less to reality but more and more to academic prestige and power.

Economic theories that fail to explain anything about individual initiative are useless in a free society, and indeed are dangerous, because they lead decision-makers away from policies which would arouse entrepreneurs to invent new products which create new jobs. None of the reductionist models allow for the power of dreams to transform human behavior, for the power of persistence to overcome obstacles, or for the power of visionaries like the Wright brothers to reshape the world. Reductionist models can predict what will happen once the free spirit of human beings has shaped the framework, but mislead when employed to calculate a new framework. That is why such models are so often wrong in our current age of discontinuity.

2. *Systematic thinking versus linear thinking*

The opportunity society examines proposals within the context of the system they will affect. Liberalism tends to take each proposal on its own and project it out in a linear manner.

Take, for example, the inter-relationship of crime, downtown business, and the chance for minority businessmen and women to be successful. Liberals are extremely committed to the last of these three components, but they do not seem to realize how much minority economic success depends on the creation of a crime-free inner city.

Because crime is very high in most inner-city neighborhoods, insurance policies are very expensive. Crime drives away higher-income people who can afford to live in safer neighborhoods, so crime restricts the market opportunities of local small businesses. Since the businesses are afflicted by a shrinking market and rising insurance rates, their profit margin shrinks, prices go up, and the circle grows more vicious: banks choose to loan money to crime-free rather than crime-infested areas because bankers fear the risk of their borrowers being driven out of business by a series of burglaries. The odds are again stacked against the local minority entrepreneur.

Finally, the high crime-rate reduces the relative prestige and status of the entrepreneur. Trying to found your own business is extremely hard work, requiring determination and perseverance. In the past, small business owners have been sustained by their relatively high social status in the ethnic community. This is no longer the case.

The Liberal tolerance of crime is linked with the economic problems of the rising minority entrepreneur. Any approach which focuses on government loans instead of making inner cities safer simply prolongs the agony.

3. *Organic life versus mechanistic life*

The welfare state uses a machine model for the process of life, a short time-frame that is appropriate for building a factory or manufacturing a commodity, but consistently too shortsighted for the slow, organic process by which people and societies evolve.

This predilection is a natural concomitant of life in the industrial world: we go home at night and flip a light-switch which triggers instant illumination; we pull the tab on a pop-top can, throw some food in a microwave oven, and turn on a television with an instant-on capability; if we have an old-fashioned television set, we may actually become impatient while we wait ten seconds for it to warm up. The degree to which we are used to everything working quickly can be measured by our anger at a car starting slowly on a cold winter's day.

Compare our modern impatience with the more tolerant attitude of a traditional farmer trying to get a mule to plow. As we enter the information age, many of its more complex systems force us to return to the more patient viewpoint of the agricultural age. Anyone who has tried to write and debug a computer program appreciates how much patience such systems require.

However, the welfare state is infatuated with the instant gratification models of the industrial society. Perhaps the essence of the welfare-state misunderstanding was contained in its misinterpretation of the Apollo project which carried Americans to the Moon on schedule. President Kennedy had challenged us to reach our celestial neighbor before the end of the decade. We met his schedule.

How often have you heard the phrase, "If we can put a man on the Moon, why can't we ——— ?" The answer is, of course, that putting a man on the Moon was a mechanical process.

The sad reality which the welfare state avoids confronting is that mechanical processes are relatively simple. Trying to provide justice for a small town of 10,000 people is an infinitely more complex and trying undertaking then getting to the Moon. Trying to measure the contributions of a dozen people to their extended family over a lifetime involves more variables and more unmeasurable quantities than did the entire Apollo project.

For most of history, men have realized that much of life is beyond mechanical calculation. Poetry, the arts, philosophy, fiction, and drama are all attempts to capture the unquantifiable. Writers like Faulkner and Hemingway, in very different ways and from very different cultural standpoints, have tried to express a rebellion against the mechanical, calculating modern world. They were right.

Faulkner's Nobel-Peace Prize acceptance line, "Mankind will endure . . . and prevail," has in it the spirit of Andrei Sakharov and Lech Walesa, of Martin Luther King, Jr., and Mother Theresa. Yet it is also a tribute to the quiet family down the street that works hard, saves every week, goes to Sunday school, and simply wants to uphold its end of the bargain we call civilization.

A good symbol of the opportunity society's commitment to an organic vision of the world is the nurturing of seed corn: seed corn is deceptively small; it takes great patience and understanding to clear the ground, plant the seeds, kill the weeds, apply fertilizer and water, and then wait patiently until the corn matures.

Liberal programs are almost always on too tight a schedule and involve short cuts which do not work. A classic example of the Liberal failure to understand organic processes is the issue of participation in government. States like New Hampshire and Vermont, in which people can have a real,impact on their communities and on decisions about their lives, have the best record of citizen participation in government. People *will* get involved locally *if* the decision process is linked to their participation in such a way that they can really take charge of their own destinies by investing their time and energy.

The problem that Liberals see in such an analysis is that, while citizens should be encouraged to participate, they can't be treated seriously because technocratic planners know more than they do. Government-union work rules and other complex agreements can't be violated simply because the local citizenry wants to have its own way. Study the fights over neighborhood control of schools in New York City: a populist uprising which would have increased parents' involvement in their children's schooling, a result Liberals would normally have applauded, was perceived as a threat to the existing power structure.

Liberalism has been trapped for two generations in a contradiction regarding this question of government decision-making versus voter participation. Liberals generally like large government structures and grandiose systems that give power to specialists; the nature of such large structures guarantees that ordinary citizens will come to feel impotent and will not participate: in the name of the people, Liberalism designed government structures that ignore people and keep them apathetic.

An opportunity society would work to develop long-term patterns and rules that gradually change society, since it is impossible to change human institutions abruptly.

4. *Traditional dreams versus instant gratification*

Nothing more clearly distinguishes the opportunity society from the welfare state than this difference in orientation toward time.

The opportunity society understands the enormous power which dreams have on people. Examples include Europeans emigrating to America, the influence of Zionism on the Jewish community in this country, and the power of freedom's lure to bring boat people through the dangerous seas between Vietnam and safety. It is clear that human beings will take great risks and endure extreme privations if their dream is large enough.

However, the effort to realize a given dream must be rooted in the historic context of the dreamers and their existing institutions. General George Marshall, for example, related part of his success in World War II to his years with the Illinois National Guard. He had learned there, he would recount in later years, that the American soldier will fight well, but he has to be given the big picture; he has to feel that he knows what is going on. That led General Marshall to develop an in-service education and information program which ensured that the American soldier could understand, to a degree unparalleled in history, why he was fighting.

Liberalism tends to ignore both the future and the past by discounting the power of great dreams to move people. For this reason, Liberalism's primary appeals to the poor are based on filling stomachs rather than arousing hopes and dreams. The message of Liberalism has been, "Vote for me and relax; I will make poverty tolerable."

The opportunity society is gambling that the chance to have your child be an astronaut or a computer programmer is more powerful than the chance to get 8 percent more in food stamps next year. Our goal is to abolish poverty by helping everyone become productive, not to keep everyone now in poverty quiet by making it tolerable.

In addition to ignoring the power of the future to shape behavior, Liberals tend to ignore the power of the past to slow down and restrict change. Beginning with Lyndon Johnson's Great Society, American Liberalism has engaged in a series of extravagant experiments which have proved the futility of trying to overpower centuries of tradition in a short time.

There was an arrogance to the Great Society which matched the arrogance of the Vietnam War. Johnson's supporters thought that America and their own management techniques were so all-powerful that they could accomplish anything. In defiance of virtually all historical experience, they actually proclaimed that we could have guns and butter too—but ended up losing both.

Consider just one side effect of the Liberal bureaucratic solution to welfare. By bringing together public housing, food stamps, and free health care, Liberals created an environment in which young girls would be rewarded for getting pregnant. The young girl who felt insignificant now had a clear avenue of advancement. The welfare state was promising to take care of her and treat her as an adult. The result has been a nightmare. White illegitimacy has skyrocketed 300 percent in the last twenty years, and in 1982 some 55 percent of all Black children were born illegitimate.

This epidemic of illegitimacy is not an accident or a sign of growing immorality among our children. People tend to do what society rewards and avoid what society punishes.

Now we discover that girls who get pregnant at thirteen don't have much education or marketable skills. The victims are not at fault for this situation. The welfare-state system's remedy for the disaster it has created is to do more of the same.

Specific steps by which an opportunity society would respond to this problem would include sending all welfare to the recipients' parents until the recipients are eighteen, thus eliminating the psychological profit of getting pregnant in order to move into your own apartment.

In addition, welfare mothers should be required either to work or to attend school full-time. This would require establishing day-care centers in public housing; but if welfare mothers did the work in these centers as a condition of their housing, food stamps, and free health care, the centers could be established at virtually no additional cost. This step would also ensure that, by the time they are twenty-five, today's pregnant teenagers would have an education, some skills, good work habits, and an employment history.

Liberals will protest that it is inappropriate to require mothers with young children to work. I find that hypocrisy to be ironic: the cost of subsidizing welfare has raised taxes enough that virtually every young suburban mother I know has worked when her children were very young.

The opportunity society will focus on creating dreams around which an energetic free people can build their lives and root our policies in the customs and institutions which have made America a melting pot, and this combination of reaching forward for hope and backward for wisdom will prove vastly more powerful than the instant gratification values of the Left.

5. *Long-term considerations versus short-term expediency*

One of the starkest contrasts between the welfare state and the opportunity society involves the question of appropriate time-frames for problem-solving. Modern Liberalism has a passion for immediate solutions. Conservatism argues that both individuals and societies have to consider the impact of expediency on the future as well as our own immediate desires.

Fiscal responsibility in the Federal budget is an excellent arena in which to view the difference between these two viewpoints. Liberals consistently argue for more spending now at any cost to the future, supporting more food stamps, more welfare—more government programs, even if the country is unwilling to pay for them. Every dollar we spend above what we raise in taxes is a dollar our children or grandchildren will have to repay. We are now so indebted that our grandchildren will have to pay both for the money we are borrowing and for the compounding interest on the money we are borrowing.

In other words, our grandchildren will have to go to work to pay for our spending before they pay taxes to support their parents and grandparents in retirement; before they pay for their own defense and their own government services; before they earn take-home pay to buy the house, car, clothing, and vacation they would like for themselves.

Liberalism is compassionate for the immediate

voter, but ruthless in its willingness to burden future voters with the cost of present enjoyment. This fiscal policy is the height of childish, self-centered gratification, but it fits the general time-frame within which Liberalism normally operates. According to Liberals, there will always be a new device to provide instant gratification and further postpone the problems—although always at compound interest.

Liberalism tends to focus on dealing with the symptom without regard to the solution's side effects. The large central bureaucracies we have accepted in return for Federal aid are an example of the solution which may have been larger than the problem. Today hundreds of minor government programs centralize power in Washington, spending an inordinate amount of money on administrative overhead while encouraging people back home to focus their attention on learning the rules and games of the Washington bureaucracy rather than solving the problems of their own neighborhood or community.

Another example of a shortsighted solution which led to a long-term problem has been our passion for keeping would-be drop-outs in school. Educational values have been subordinated to our willingness to make schooling attractive to disinterested students, thereby creating discipline problems in the classroom, boring the brighter students, lowering the value of a high school diploma, and permitting undereducated students to receive diplomas which they cannot even read.

The myopic focus on the immediate problem of teenagers who did not want to learn the values and knowledge of society ensured their physical presence in school but did nothing for the larger issue of convincing them to value education as a source of income, power, and wisdom.

A program which permitted people to drop out of high school and get menial jobs to learn that ignorance is expensive would have a greater effect on young people both in school and out. Such a program would acknowledge that learning is indeed of value but that it is also difficult. This would make schools prestigious places to be. Drop-outs could be allowed to return in later years when they are ready to learn.

Adults have an obligation to consider matters in context; they must be aware of their obligations to others. Adult politics must include a willingness to make the decisions which work best in the long run. We must have a commitment to creating a better future and not just solutions to immediate crises.

This style of adult politics and adult government will be complex, requiring more patience than recent Liberalism has displayed. Mature individuals learn that easy symbols do not earn paychecks, pay mortgages, and raise children. Real life requires a complex and adult approach to life's problems, trying to find real solutions that will last. The same is true in politics and government.

6. *Geopolitical realities versus the arrogance of a self-centered America*

President Kennedy's promise, in his inaugural address, that "We shall pay any price, bear any burden . . . to assure the suvival and the success of liberty," was an example of a dangerously overblown sense of self-esteem and national power. In fact, of course, we were not willing to go to these extremes to protect liberty. We failed in Vietnam to meet the challenge Kennedy outlined.

David Halberstam's *The Best and the Brightest* underscores the point that the intellectual managers whom Kennedy assembled were both overly impressed with the power of their own ideas and content to remain ignorant of the tenacious nature of world problems, whether those problems involved the North Vietnamese in Asia or poverty in the South Bronx.

The power, complexity, and instability of the geopolitical universe is beyond our comprehension. Just as good sailors learn to respect the power of the ocean, so good politicians learn to respect the capacity of the world to overwhelm their best laid plans.

Real mastery in any art form comes through subordination of your personality to the objective requirements of your trade; then, having mastered the trade, you must allow your personality to guide your skill; if your personality dominates your skills, you will rapidly lose your mastery. Writers who have a best-seller sometimes begin to think that their personalities rather than their skills brought them success: presently they learn that arrogance without discipline is a sure path to failure.

Our society remains overextended despite the lessons of the past fifteen years. We wish to be powerful in every part of the world, to compete in every industry, to educate every child equally, to ensure everyone's equal access to the best health care on the planet. We have attempted a whole series of simultaneous projects as though our human and material resources were limitless.

In fact, trying everything guarantees failing at everything. As Frederick the Great warned, "He who defends everything defends nothing."

The first lesson to be learned from the world's size and complexity is that we must set priorities. We must realize that our resources are frighteningly limited in comparison to our dreams and obligations. In foreign policy, we must insist that our allies bear their fair share of the burdens of defense. We must be very cautious about where and when we commit ourselves economically or militarily. To protect our diplomatic prestige, we must weigh each potential obligation against the obligations we already bear.

At home, we must liberate people from bureaucracies so that they can achieve more for themselves. Even in a more dynamic America, our resources will be at a premium. Whatever the long-term future, for the next generation we will have to pinch pennies to balance our national budget and bring our national debt under control. Only then can we bequeath our grandchildren a country in which they can spend the wealth they create.

Second, since the world is so complex, we should expect to be surprised regularly. Much of recent American history has been a sequence of surprises:

from abroad, we have reacted with shock to the Tet offensive in Vietnam, the Soviet invasion of Afghanistan, the hostage crisis in Iran, and the truck-bomb attacks in Lebanon; at home, we have been taken unawares by waste in the Pentagon, sex scandals in Congress, children using drugs and alcohol, traditional jobs becoming obsolete, and farmers suffering through droughts. None of these surprises should have been as great as they were.

I was first struck by this American passion for avoiding the lessons of history when I watched the Atlanta real estate boom of the early 1970s. In the race for wealth, many companies brought in young real estate developers, specialists who applied Harvard Business School techniques and used the finest mathematical analyses to maximize short-term profits. I asked some of these people in 1972 and early 1973 whether they had ever heard of the Florida land boom of the 1920s. None of these so-called experts had ever delved into the historical cycles of their industry. They were just technicians.

The bust came in 1974 when an overpriced, overexpanded market collided with an interest-rate crunch. The go-go developers went. Not only did they bankrupt their own companies; they endangered many major institutions—including the largest bank in Georgia, which had foolishly allowed itself to believe that a new age of permanent prosperity had arrived.

There is nothing new about real estate busts. What was new was a generation that thought it was big enough and powerful enough to impose its own rules on reality.

The opportunity society would manage its plans and programs within the realities of the world at large. If we are moving into an information age, then the opportunity society would move us there quickly. If we must scale down jobs in many heavy industries, then the opportunity society would encourage workers to move from old industries to new ones.

However, since we know that we are fallible, we would be cautious about building a central bureaucracy to direct most of this transition: even when we have a vision of where we must go, government is almost never the right mechanism to carry us there directly. Because the world *is* so difficult to know, markets and decentralized initative are the best mechanisms for exploring reality and responding to changes. Thus, an opportunity society would guide the decentralization of American society to make it easier for us to supersede our large institutions.

We have to learn that the world has objective rules which have worked throughout recorded history and which continue to work. For the last twenty years we have violated these rules, and we are paying substantial prices for those violations.

For example, every successful military force has had to learn to delegate authority to subordinates so that those who possess great fighting qualities can make their own mistakes, garner seasoning, be identified and promoted. Yet our modern forces have used the new communications systems to permit succeeding levels of command to overcontrol their subordinates. We have seen President Kennedy talking directly to destroyer captains during the Cuban missile crisis or Lyndon Johnson ordering specific air raids and being awakened at night to be told how they were doing; we have had ample time to regret Carter's mismanagement of the antiterrorist force during the abortive Iranian rescue attempt and mourn its casualties. In some engagements in Vietnam, a company commander had above him his batallion, brigade, and division commanders, each in his own helicopter, each interfering and confusing the battle. This is a professional military violation of the first order.

Not only must we deal with new abuses brought about by the misuse of new technology, but now we find ourselves re-experiencing the problems of the past. Work in our bureaucracies *does* tend to expand to fill the time alloted, as Parkinson predicted. Bureaucrats *do* tend to become territorial and self-serving, as Adam Smith warned us. We must recall that, as humans and as a nation, our fates are linked;

we must stop breaking moral rules before corruption becomes an unbreakable rule.

We will never move on to real solutions to our current problems until we humble ourselves as a people and relearn the basics of human interactions. Though technology changes, human beings remain remarkably unchanged—as any comparison of contemporary vices and the Old Testament could show. For all our efforts, human beings have not invented one new form of sexual deviancy in three thousand years. If, as a species, we have not invented anything new in vice, why would we think we are likely to invent something new in virtue? "New technology, old commandments," is the starting point for an opportunity-society approach to building on reality.

7. *Citizen emphasis versus bureaucratic emphasis*

If our free society is to remain healthy and free, it must constantly stimulate and strenghten its citizens by encouraging them to relearn the habits and skills of governing themselves.

There are many habits and attitudes to cultivate in order to achieve successful self-government: we must be passionate enough to argue for our dreams and our beliefs, but tolerant of others arguing against our views and for very different dreams and beliefs.

We must acquire the toughness of political campaigning and debate while maintaining the good fellowship of citizens engaged in the noble quest of self-government.

We must delegate power to our representatives and senators, to our President and our ambassadors and generals, while remembering that, in a free society, we each have the moral responsibility for the actions of our government.

Self-government is so complicated and so demanding that it can only be learned by doing: it is the action of governing ourselves which teaches us to govern ouselves.

Since self-government leans so heavily on the skills, intelligence, and energy of the people, it is vital that we adopt policies which encourage self-reliance, involvement, and a spirit of independent toughness.

The virtues necessary in a free, self-governing society are contradictory to those required in a centralized bureaucratic state: the meek, obedient, subordinated member of the central bureaucratic system does not have the strongly independent personality of a true free person.

Independence and self-reliance run very deep in our culture, and they are surviving despite the best efforts of the welfare state to destroy them. If we obeyed every bureaucracy, implemented every Washington-invented plan, adhered to every Washington politician's centralized vision, we would rapidly become a country incapable of sustaining our freedoms.

Every step toward centralizing decisions in Washington undermines local opportunities to practice freedom.

An opportunity society would place greater value on the practice of self-government. Whenever possible, the opportunity society would structure the processes of government so that each citizen could be involved in reshaping the process of administration and returning to the people the power of decision over their lives.

An opportunity society would also emphasize individual self-reliance. Every effort would be made to balance the Federal budget so that individual citizens could borrow money to buy houses or cars or to start businesses and farms at reasonable interest rates.

An opportunity society would be committed to ensuring that all citizens have power over their daily lives, have a healthy control over the bureaucracies which serve them, and have the personal resources in their take-home pay and their savings to behave as their own masters. This is truly power to the people.

8. *Personal ethics versus governmentally enforced rules*

The welfare state's antagonism to traditional values and to the central role religion plays has been eroding free society.

The ACLU's anti-religious model is unrelated to the beliefs of our founding fathers, who began their meetings with prayers and who routinely asked God to bless their struggle for independence. Indeed, the roots of the rebellious tradition from the 16th-century Dutch uprising, through the 17th-century English civil war, to the 18th-century American Revolution, are to be found in religious fervor and religious commitment.

The great struggle in Jefferson's day was between freedom for religious differences and the imposition of an established religion. Given that choice, the founding fathers chose diversity. However, none of them argued for freedom *from* religion. Jefferson—normally thought of as one of the most radical of the revolutionaries—directed that students at the University of Virginia could have the freedom to choose which religious service they would attend. He did not offer them the freedom to avoid religious education.

The founding fathers themselves had proven what out-numbered, out-gunned, and out-financed men could do if their spirit was strong and their moral commitment prepared them to endure hardship. They, of all men, were the least likely to neglect the importance of religion.

An opportunity society would restore voluntary prayer in school, re-emphasize the importance of ethical and honorable behavior, more rigorously punish dishonest behavior, and do everything possible to instill in its citizens the importance of spiritual commitment. There is no other path to the preservation of freedom.

Here, then, are the key points on building an opportunity society:

1. A free society must have vision so that people can organize their lives around opportunities that will help both them and their country.

2. An opportunity society must be organized around the coalitions of the future in order to compete with a welfare state built around the coalitions of the past.

3. We must set idealistic goals and put them forth in a fashion that will motivate but not frighten people about the inevitable transition ahead—a transition to a better life.

4. The government that promulgated the Homestead Act, offered subsidies for the Transcontinental Railroad, and developed the land-grant college system to modernize farming must reassert its rightful role as developer of incentives for change. Conservatives who resist this active government role must be reminded that you cannot keep Panama Canals unless you first build them.

5. We must expedite the transition from a welfare state to an opportunity society wherever possible by developing positive incentives. We must leave it to the Liberals to attempt to change people through punishment. Whenever possible, we should reward people for making changes in their lives that will keep the country profit from speeding up the transformation to a new era.

6. Citizens must be seen as resources capable of helping themselves and each other, rather than as victims to be protected by the bureaucratic state.

7. While government can build or encourage systems whenever possible, these systems should be nonbureaucratic and designed so that individuals, communities, and collective enterprises use them. The Homestead Act, through which one land agent per county administered a program that allowed millions of settlers to open the American West, should be our model.

8. We must focus responsibility for change on those who will profit from the change and, whenever possible, use individual citizens as catalysts in areas where they will profit. Given enough hope for a better personal future, people will find the energy, the time, the persistence, and the intelligence to change the world.

9. The opportunity society must be a magnet which attracts every American who is excited by hope and opportunity. It is our goal to replace the current compartmentalization of the welfare state with a

new visionary politics. Our new coalition is as encompassing as F.D.R.'s New Deal, as open to every American as Jeffersonian and Jacksonian democracy, and as dynamic as Teddy Roosevelt's vision of an America that could lead the world.

10. We must tell people that hard work and persistence are the keys to success in a free society. The fact that we can't deliver miracles tomorrow is less important than the fact that we can begin working together today to create miracles as soon as possible. There will be Liberal pie-in-the-sky promises and Liberal smear campaigns which we must counter with adult proposals and with a campaign of truth that, upon analysis, is more believable than the smear campaigns.

11. Whenever possible, we must develop governmental services and programs at the least centralized and most human level possible. Instead of controlling people through big government, we must make government small enough that people can control it.

12. In almost every field, we must focus on reigniting the spirit of free enterprise which once made this country so dynamic. The key to a healthy, free society is the willingness of thousands of ordinary Americans to realize their dreams.

CHAPTER 6: REFORMING THE GREAT BUREAUCRACIES

America today is in thrall to seven major bureaucratic structures, all of them expensive and personally burdensome to the public—structures whose entrenched power to protects themselves from change slows America's progress away from the welfare state toward an opportunity society.

These entrenched structures require real and innovative reform. They are:

1. The legal system.
2. Health care.
3. Education.
4. Welfare.
5. Taxation.
6. Public bureaucracies.
7. The Department of Defense.

These areas—all funded largely by the taxpayer, but dominated by a professional elite whose voice serves only those professionals—share certain characteristics.

Each structure's insiders regard themselves as an aristocracy serving society in a way which people outside their specialty can't really understand or appreciate. Whenever one interviews a medical doctor, an attorney, a classroom teacher, a welfare administrator, a tax collector, a civil servant in a large agency, or a professional soldier, one is continually reminded that the professionals have special knowledge and their jobs have requirements which the rest of us can't really appreciate.

In each area, the structure has now grown so large that most of the professionals spend their time talking only to other professionals. At a seminar in any of these fields, there will be at least some panels on topics so obscure that only specialists will be able to understand the ideas being discussed.

The professionals ruling each structure have associations which guard against threats from the outside, providing staffers for congressional committees, specialized magazines, academics studying topics, executive branch agencies, and lobbying groups.

Each professional association already has a clear vision of its function: to accept as legitimate only those tactical and operational reforms which fit into their existing vision and strategy. In each case, the basic framework of thought for the current system is at least forty years old and the models of bureaucracy, career enhancement, and prestige date back to 1950 or earlier.

These large, self-protecting systems value their professional concerns far more than the public interest. Any proposals for substantive reform by "outsiders" are derided as the views of amateurs who fail to grasp the sophisticated problems of these professionals.

In an age of change, such antiquated systems are expensive and inefficient, yet the inherent capacity of each structure to resist reform has prevented the growth of new techniques within each specialty—reforms that might benefit the public.

The professional elites have established a mystique of "higher values" which overrides the market questions of "Can it be done for less?" and "Is it really what people want?" and which allows insiders to fix reasonable costs and appropriate processes, thereby determining what the product will be.

Finally, each structure is ruled primarily by the internal needs of its professional elite, whose response to any public request for improvement is to demand more money. The structures themselves never offer reforms which would achieve better results without an increase in cost.

An example of a structure's ability to expand its share of resources has been Supreme Court Chief Justice Warren Burger's crusade for more Federal judges. Any layman who has looked at the anachronistic court procedures, the inefficient management style, and the petty arrogance of many Federal judges would conclude that court reform requires streamlined procedures and administration to bring the judicial system into the 1980s.

Instead of reform, Chief Justice Burger requests more judges, who would need more courtrooms and more administrative assistants—all of which require more money and continue the present anachronistic, time-wasting system. Chief Justice Burger's willingness to charge the public more so that the legal profession can avoid reforming its procedures is typical of all the major professional systems.

Health professionals suggest that better health requires more money. As a result, health care has increased its share of the gross national product from about 5 percent in 1950 to over 10 percent today. The professional answer continues to be more money—either for the doctors, if you are a typical Conservative; or for bureaucrats who will control the doctors, if you are a Liberal.

Similarly, the National Education Association will explain that better education requires more money and that this money should come from the Federal government (which has a deficit of hundreds of billions of dollars) because the state and local government (which are generally balancing their budgets) do not have the extra money.

Our welfare system is destroying families, encouraging an epidemic of illegitimate births, keeping young women in poverty, and accustoming children to living on welfare. Nonetheless, our welfare professionals suggest expanding the same system which created the current morass.

When the complexity of our tax system leads people to cheat, Internal Revenue Service officials write additional rules and do additional audits which cause people to feel that defiance of the system is morally justified.

Our general bureaucracy functions in obsolete ways, using public administration systems of the 1920s and office equipment of the 1950s. Public administration unions claim the answer is to raise the pay of the bureaucrats and to hire more people from a roster compiled on the basis of factors other than performance.

Finally, our military hardware is so expensive that we don't have enough of it and our manpower costs are so high that we can't afford sufficient troops to meet our obligations. The defense lobby demands we pay even more money to giant corporations which produce expensive equipment to be used by the most expensive military we have ever fielded. Meanwhile, military pensions threaten to exceed pay for active-duty soldiers.

The problems of the seven major bureaucratic structures are similar, but the potential solutions are sufficiently different that we must examine each on its own.

A Fair, Inexpensive Legal System

The United States is today the most lawyered and the most litigious society in the world. There are more lawyers in Georgia (population: 5,500,000) than there are in Japan (population: 100,000,000).

Despite our wealth of attorneys, most Americans would agree we have a dearth of justice: criminal justice is too slow, too lenient, and too random; civil actions are too expensive and too protracted.

We must de-emphasize courtroom activity and expand the use of mediators and arbitrators who need not be lawyers. In most cases, law should be simple with a more sophisticated, more professional style of law for complex matters.

A legal system which met the needs of ordinary citizens would differ markedly from the current model. First, the criminal justice system should make life easier for victims and witnesses. After reasonable notice, the burden of being prepared and being in court should be on the accused and the defense attorney. Those attorneys who are particularly disruptive or exploitative of the system should suffer economic and, ultimately, professional penalties for having failed to behave in a manner beneficial to society.

Next, we should excise the legal fees now built into the American business system which raise our costs and make us less competitive overseas because of the overhead earmarked for possible litigation.

Finally, we must simplify the intricate processes of the legal profession which have so infused government that the bureaucracy increasingly uses language and patterns which are quasi-legal. The result is additional paperwork, complicated formulas and procedures, and increasingly complicated systems for adjudicating claims.

Today, we start with the attorney's habits and preconceptions, then pass on to our citizens the cost of living in a nation of crowded court dockets, frequent lawsuits, and incomprehensible government regulations. We need to redesign our legal system around the citizen.

The Best Possible Health Care For Everyone

Health care, today the most expensive single factor in American life, takes twice the share of our gross national product as does national defense. The cost of health care is increasing more rapidly than any other element in our economy.

People want to live. In our free society, citizens exert great pressure through their political leaders to ensure access to the medicine and the doctors necessary to stave off death. Our greatest success in health care in the last two hundred years has not been in better hospitals or better doctors, however: the real advances have come from eliminating the epidemic diseases which once took so many lives—cholera, smallpox, malaria, yellow fever, typhus, typhoid, and bubonic plague. We now live in a fundamentally new era in human history because of the eradication of the microparasites which used to be endemic.

The central problem with our approach to the need for adequate health care at reasonable cost has been its focus on tactical, or at best operational, changes within the existing vision and strategies of health care. An example of this visionary change is the elimination of polio as a major threat. Only thirty years ago, families worried every summer about the inevitable outbreak of polio; children who were faced with life in an iron lung hoped for improvements in iron-lung technology; rehabilitation of the stricken was a long and difficult road at best. Jonas Salk's development of a polio vaccine represented a paradigmatic shift: inventing a better iron lung machine would have been a significant tactical breakthrough, but it would not have altered the world; the discovery of a polio vaccine was a response to a vision so great that it did indeed alter the world.

There are eight specific problem areas which need redirection. If we do not implement this redirection, the inexorably rising costs of the current system will lead to a bureaucratized, red-tape dominated, welfare-state model of socialized medicine.

As you examine these areas, remember that the sheer cost of the current system is going to force even Conservative administrations to implement Federal government controls when the implications of the projected costs for Medicare and other health programs are understood. There will be *change* in health care—either toward centralized control and payment or toward the visionary innovations of the opportunity society.

The eight areas of necessary change are:

1. Focusing on preventive medicine and good health.
2. Increasing research at the vision and strategy level to transform specific areas.
3. Decentralizing decision-making so we can move more rapidly to new technologies and procedures.
4. Sharing new medical knowledge more widely and systematically so people can adopt new methods without professional involvement.
5. Relying more on the patients and their families, friends, and neighbors, instead of on health-care professionals.

6. Returning power and responsibility to the patients and the local institutions.
7. Eliminating national responsibility for what ultimately must be individual decisions.
8. Changing the third-party payment system so that the individual is once again directly involved in the market costs.

Taken together, these steps would revolutionize our current health-care bureaucracy by providing a more flexible, technologically-advanced health-care system, which would cost less and involve less Federal interference. Our goal must be a progressive system of health care which will evolve as rapidly as possible.

Virtually every centralized health-care system has tried to restrict health care to the technology and processes of the previous generation. Bureaucrats and politicians, rather than doctors and medical specialists, make the decisions in a socialized health-care system; the resultant financial crunch leads to decaying hospitals, obsolete equipment, and inordinate pressures on doctors.

A recent Democratic proposal called for jailing doctors for six months if they failed to obey new government regulations on accepting Medicare patients. This is an example of the arrogance of big government at its worst: having designed a Medicare system so bureaucratic that people didn't want to work with it, the bureaucracy wished to jail people for failing to comply with its voluminous regulations.

Yet the underlying crisis of health care is real and must be solved, even if the Democratic approach of jailing doctors is not the answer. We cannot permit massive costs to continue doubling in succeeding generations.

Faced with skyrocketing costs which threaten to bankrupt the Medicare fund, the Reagan Administration is proposing a series of national reforms, involving more bureaucracy and more paperwork, that will try to impose change from the very top with a series of prospective payment systems. Hospitals will soon be arguing about the formulas under which they should be paid; doctors will be sharing insights into how best to do what is medically necessary despite Washington regulations.

The medical community has failed to develop a less expensive approach to providing health care for everyone. The Reagan Administration has failed to understand that any national bureaucratic response—even the best-intentioned—is doomed to failure.

Any real solution must combine a new approach within the medical community that will lead to less expensive care with a government willing to decentralize and debureaucratize the current system of health care. Let us therefore look in detail at steps which could change our approach to providing health care.

1. A new focus on preventive medicine and good health would change the entire direction of managing health care and agree with three opportunity-society principles: first, our goals should always be positive rather than negative; second, we are interested in general, rather than partial, solutions; third, we involve and strengthen the individual citizens, rather than focusing on them as victims.

We must convince people to watch their diets and habits. We may want to reward people who do not need health care. There have been some experiments (the Mendocino, California, teachers' association is one) in which people are given money at the end of the year for not using their health-care plan.

For instance, we could simply offer a $500 year-end bonus to people who have not used Medicare in that year. Citizens who needed Medicare would have it available without the out-of-pocket expense of a deductible cost higher than the current $50. In effect, though, it would become worthwhile for citizens to pay up to $500 on their own, instead of using the Medicare system. The long-term effect of rewarding people for watching their health, taking care of themselves, and being responsible for minor health costs would more than pay for the year-end bonus.

Similarly, we must bring physical education teachers, recreation program directors, and health spa staffers into the process of developing sound public health. Every citizen who avoids a heart attack saves the community a lot of money. Now that we know people will be living into their eighties and nineties, it is greatly to society's advantage to encourage them to take care of themselves. That way they can be physically and mentally active all their lives. Every month that people can live in their own homes because they are physically and mentally healthy saves hundreds of dollars in nursing home bills. It would be worthwhile to encourage people to participate in some physical fitness program, especially when they are over forty.

Finally, we should build a much more direct relationship between bad health habits and the cost of those habits to society. A health tax on the alcohol and tobacco industries would begin to recoup the costs they impose on society at large. At present, non-drinking taxpayers pay for the problems of alcoholism through public health clinics and other programs, and non-smokers subsidize the hospital costs of lung cancer. We can't define *all* the ill health caused by these two major voluntary epidemics of our time, but it is possible to gain some general assessment of what they cost the taxpayers. Besides the revenue, there would be two beneficial side effects from charging alcohol and tobacco companies for the costs of the lifestyles to which they pander: we would marginally discourage alcohol and tobacco abuse, and we would encourage people to watch what they drink and smoke. Today we economically encourage the alcoholic and the potential lung cancer victim by asking all the other taxpayers to subsidize the costs of these addictions.

People can become aware that their habits affect their health. People can learn to view the doctors, the hospitals, the health spas, and recreation departments as their allies in the effort to be healthy all their lives. When they do, we will have a more positive nation, more active senior citizens, and a less expensive health-care system.

2. There are a number of definable problems which are extremely expensive and which deserve major investments to solve. The classic example of a good idea grown very expensive is the kidney dialysis center. Kidney dialysis allows thousands of Americans to live much longer. Dialysis centers have expensive equipment which is operated by well-trained, well-paid people.

The more people we keep alive through kidney dialysis, the more centers we need. The more centers we need, the more expensive the program becomes. We are currently spending over a billion dollars a year just on kidney dialysis.

The welfare state response to the rising cost of kidney dialysis will be to accuse the centers of gouging the public, to establish tighter guidelines, and to hire more bureaucrats to audit more centers. There will presently be horror stories about some investors and some doctors making money from kidney dialysis. The government will squeeze payments and fight against future cost increases. An association of kidney patients and their families will be formed to pressure their Congressmen and Senators for money to save their lives.

A more fruitful opportunity-society strategy would be to offer a very large reward, possibly $100,000,000, to the inventor or inventors who can develop an inexpensive, self-applied kidney dialysis system. Our intended impact should be that of the polio vaccines which eliminated iron-lung wards in hospitals. We want to make it so simple for kidney patients to take care of themselves that, a decade from now, the kidney dialysis center has become a relic of the past.

When you reflect that kidney dialysis will cost well over ten billion dollars in the next decade (and may cost nearly fifty billion dollars between now and the time our grandchildren are of voting age), then offering 1 percent of that amount now as a prize is a wise investment. There are already companies and researchers interested in this problem. With a $100,000,000 reward to encourage their efforts, we may well solve the problem within three or four years. That would save literally tens of billions of dollars before the end of the century.

The welfare state has no interest in this sort of visionary research to change the entire framework of health care. Certainly, it could not accept the concept of $100,000,000 rewards for such research. The medical community now despairs of defending the idea of profits, so it would not think such an offer practical or possible.

We need to assess those areas of health care whose current or probable future costs are sufficiently large that they justify either an expanded research investment or a reward system. For particularly expensive areas, we should follow both tracks in parallel.

Recurring annual costs are useless for the purpose of really visionary management. In considering the value of major breakthroughs, we need to analyze health-care costs on either ten-year or generational scales. Then we can begin to assess our need for research to develop less expensive approaches.

This plan offers a world of considerable challenge and change for the health-care profession, some of it no doubt distressing. But it offers the public an opportunity to thwart the welfare state's tendency to hinder change through red tape and bureaucratic over-regulation by replacing it with an opportunity-society transformation utilizing new technology and new processes.

3. In order to change the health-care system rapidly enough, we must return to a more decentralized decision-making process. Peters and Waterman, in their study of management success, argue that the big companies who do best are those which pretend to be small, decentralizing so that many little decisions can be made quickly and at minimum cost. Peters and Waterman also emphasize that the company which has a new idea and gives it to a few people for experimentation will consistently outperform the company which takes the same idea and assembles a large staff to analyze it thoroughly.

Clearly, the history of successful innovation leads

to a model in precise opposition to that of the centralized bureaucracies which increasingly characterize health care. In the government's Health Care Financing Administration. In the private Blue Cross/Blue Shield group, and in corporations whose hospitals are spread across the nation and the world, too many decisions are made by a few senior officers. Such a system lends itself to defense of the past and glacial resistance to change.

We need to make it relatively easy for local doctors, hospitals, and communities to experiment. We need to encourage diversity in health care so that we can have hundreds of alternatives across the country.

We also need to reward innovation in health care. Today Medicare and its parent, the Health Care Financing Administration, are essentially negative institutions which punish those who make mistakes but ignore innovators who might revolutionize the industry.

Our health-care system should encourage procedures and technologies which allow patients to be tended in the office. It is in our interest as taxpayers to overpay for specific procedures if, as a result, we save money over the total cost of the alternative. Our analysis must consider the *entire* medical process and not merely the accounting sub-unit with which a particular bureaucracy deals.

4. We must get new technology into the hands of the general public as rapidly as possible. Many of the procedures currently available could be used by laymen. New software for home computers will allow an individual to become proficient with relatively little training. Videotape systems are already being used by some hospitals to train patients to know what recuperation schedule they should expect when they leave the hospital after a particular operation.

It is in everyone's interest that individuals do for themselves as much as possible; when they cannot, a large, well-trained corps of paramedics should be available; eventually, only the most sophisticated and complex procedures should have to be done by top professionals. In the long run, no society can afford to pay professional salaries and fees for work which requires only semiprofessional expertise.

As an example, the trend toward bringing midwifery back into fashion is a good start toward delegating responsibility. There are many procedures analogous to midwifery that, if encouraged, would lead to better-informed, more self-reliant citizens, and to a more appropriate structure of health care.

5. We cannot afford to let the welfare state's vision of the citizen as victim combine with its addiction to professional bureaucracies. Teaching people to be content in helplessness weakens them and leads them to doubt their ability to care for themselves. The welfare state risks a downward spiral in which people form a dependency on professionals whom they can ill afford.

Hospital stays and nursing home admissions provide two examples of this excessive dependence on professionals and on professional institutions. The home health-care industry has been developing a number of systems and processes which allow people to take care of themselves or to be taken care of by their friends and relatives. The current structure of health-care reimbursement, however, is at best neutral and at worst actively hostile to these efforts to get people out of hospitals and nursing homes.

Home-computer monitoring systems, tied to a telephone and radio-connected to emergency squads, may provide liberating flexibility for people convalescing or rehabilitating themselves at home.

"Lifeline" is a wrist radio about the size of a wristwatch. If wearers fall or are hurt, the push of a button on the Lifeline unit activates their telephone, summoning an emergency team which shows up within three to five minutes. For $500 per installation and a $12-per-month maintenance fee, senior citizens can stay in their own homes with Lifeline instead of having to go into a nursing home. The first two months of the average nursing home's fees will pay for the installation of the system and the first four years of its use.

It is important to remember that the current Lifeline wrist-radio-telephone system is in its technological infancy, equivalent to the pre-Model T era for automobiles. But its potential deserves research and development.

We need, in fact, a series of research and investment programs to develop home health care, home maintenance and communications, and individual training. These ideas would revolutionize our current hospital and nursing-home systems.

We also need incentives to encourage people to get their friends, neighbors, and relatives to help take care of them. Today, if you stay in a hospital, the government will reimburse you. If you go home and your son or daughter takes care of you, we offer no incentive. It would be worthwhile to develop a system wherein your family could earn a tax credit, or even a direct payment from the government, for every day which the doctor certifies you would have been hospitalized, had not your family taken care of you. The government could save $277 a day per patient by encouraging home care. Your family could learn more about medicine and health care. You could be in warmer, more human surroundings.

Shouldn't it be possible to design a system to let that happen?

6. We will never enjoy a responsive, rapidly improving health-care system so long as most of the key decisions are made by large bureaucracies. The size of the Health Care Financing Agency, the Medicare system, Blue Cross/Blue Shield, and similar systems makes them unwieldy and prey to internal politicking.

We must return both power and responsibility to local communities, to patients, and to the professionals. A major savings in health care could be made by simply eliminating large portions of the administrative overhead now in place. There are over 200,000 people shuffling paper in the insurance industry and in state and Federal government—people who never see a patient, never build a machine, dispense medicine, or investigate a disease; yet they cost us $12.6 billion per year.

In addition to the direct cost of external overhead administration in health care, there is an additional cost: each doctor and hospital now must hire people who never see patients but simply shuffle more papers to match those being shuffled in the insurance, state, and Federal bureaucracies. The average cost of this additional layer of personnel is four cents of every dollar spent.

Thus, we, as citizens and patients, are paying at least thirteen billion dollars of additional money for purely administrative overhead—a "necessary" cost of the way the welfare state has designed the current system. A more decentralized, more locally responsible system would have these dollars to invest in research and in better services—even to waste on mistakes.

I specifically include the option of wasting money on mistakes because it is less expensive and more productive in the long run to make mistakes locally rather than to build a centralized system large enough to prevent mistakes. The costs of auditing and control can be far greater than the cost of a more relaxed, sloppier, more localized structure. An opportunity society would willingly tolerate a certain level of waste and duplication because the cost of avoiding it would be even more expensive.

7. We must return to a local-community focus on providing health care and developing good health.

Once upon a time, communities had to raise their own money and to think seriously about how to spend that money: if they wanted the latest X-ray machine, they had to figure our how to buy it; if an expensive piece of equipment was not utilized, then the community suffered the consequences and was more cautious the next time its medical staff came up with a good idea.

Once the Federal government got in the act, everyone wanted to buy everything because no one had the responsibility of paying for it. Local communities had no incentive *not* to buy every new machine that was offered. To slow the cost increases, the government undertook programs to supervise the process of health provision in order to restore order and efficiency.

The Health Systems Agencies, classic examples of the welfare state focusing on a symptom and creating a new problem, were the result.

The Health Systems Agencies were intended to represent the public interest by developing a health-care plan for a large area. Major decisions on new beds or new technology had to come through politically appointed, relatively uninformed committees.

A trio of predictable disasters occurred.

First, the committees were rapidly politicized: shrewd professionals learned to play the political games necessary to get their way; hospital administrators boasted openly of how they wined, dined, and maneuvered to get what they wanted. We had diverted emphasis from the local community's needs to the political skills of the local community's health representative.

Second, the committees had a bias in favor of the past. Efforts to open hospital beds in high-growth suburbs were met with suggestions that the region was overbedded since older, downtown hospitals in areas of declining population had too many beds. The people who staffed and really dominated the agencies lived in districts near good hospitals and tended to be Liberals who liked big city environments, who thought that suburbs and small towns were gauche, and who reinforced this bias against new hospital beds in politically "wrong" areas which needed them.

Third, since the welfare-state politicians and bureaucrats were appointing people to the agency, there was a bias against technology as well as a sense that non-profit agencies were good, but that people who were out for a profit were to be distrusted. Many of the breakthroughs which will provide better health care at lower cost involve higher technology, and such technology comes first into the hands of doctors willing to risk their own capital on new equipment which may ultimately make them money. The Health System Agency therefore discouraged behavior which was likely to make the system run better.

Consider, for example, the track record of health systems agencies and CAT scanners. CAT (computer assisted tomography) scanners are the great analytical breathrough of the last twenty years. They are for our generation what X-rays were for our grandparents.

First-generation CAT scanners are very expensive. In order to keep down costs, the Health Systems Agencies restricted the purchase of the equipment to just a few hospitals. In a welfare state, that sort of bureaucratic rationing is considered positive and useful.

This saves money, but it makes no sense: prices will stay high as long as only a few machines are built; when the market is big enough, several manufacturers will compete to make money by building them, the price will crash, and CAT scanners will become more affordable. Furthermore, the proliferation of scanners will guarantee that are there are repairmen who know how to maintain them. Finally, widespread use of CAT scanners will expand the quality of health care by replacing current surgical explorations, thus saving both lives and money.

If you look at the introduction of the X-ray or the development of virtually any other major technology in 20th-century America, you will find that it involved the kind of price crash where I have posited for CAT scanners.

The lessons a good bureaucrat learns are precisely the opposite of those necessary to manage change. For this reason, we must debureaucratize health care. If we hope to continue to lead the world in inventing new systems, medicines, machines, and processes for staying healthy, then we must go back to a system in which local communities take on the responsibility of risk-taking. We have to find ways for them to bear more of the economic burden locally—and for them to keep more of the potential profit. Thus, we must break down the current central bureaucracies, including the centralized payment systems.

Doctors and hospital administrators who hate Medicare rules but love Medicare checks can't have it both ways. If the bulk of the money is going to come from big insurance companies and big government, then there will inevitably be big red tape and big bureaucracy to follow. The money has to be protected in Washington if it is going to come from Washington.

We, as a society and as citizens, can and should insist that all future health-care reforms involve less central government. We should begin now to develop new local-option approaches which would allow doctors, hospitals, and communities to contract for services. If a community wants to improve Medicare in its county with a better system, we should allow it. An opportunity society must encourage local options, inventions, and experiments.

8. We must reform the third-party payment system, which ensures a high level of irresponsibility by both the patient and the provider because neither has any direct concern for cost. Only a welfare state would expect people to restrain themselves instead of exploiting a system designed for exploitation. In fact, precisely the opposite happens: people feel that, because they have already paid for their insurance, the insurance company owes them the most expensive health care, even if it is not necessary.

Originally, the insurance system was set up so that you could put money aside on good days to provide for adequate health care on bad days. When health insurance became a tax-deductible cost, businesses provided it as a fringe benefit without too much concern for cost—the insurance companies simply wanted the premiums.

Now that the cost of health insurance has become a major factor in the cost of living in America, businesses collectively pay $65.7 billion toward health insurance for employees. Those are dollars that, because they are not available for take-home pay or for capital investment in new factories and tools, add to the cost of American products and make us less competitive in the world market.

We face the same general problem as a nation: the current projections are that, by the year 2000, Medicare will cost $332 billion, requiring an average worker to work 1.3 weeks just to pay for Medicare.

The key to improving this system is to find methods of collective insurance which provide incentives for the right behavior.

One step might be an individual health-care account in which you would buy a catastrophic health-care plan reinsured by the government. Such a plan would ensure that unique illnesses or family tragedies would not cause people to lose their homes or go bankrupt. All of us are vulnerable to disaster, and only all of us together can afford protection against this vulnerability.

But catastrophic illnesses are only a minor part of the health-care picture. We can afford catastrophe insurance for everyone—*if* we can get ordinary costs under control.

The second part of the individual health-care account would be a system much like an Individual Retirement Account, involving some co-insurance in which you bought coverage against major illnesses which were not catastrophic, but which cost more than you could afford to spend out of pocket.

Thus, you would be protected against any major illness, first by your insurance company and then by your government. For the bulk of minor illnesses, however, you would be drawing against your own money every time you used your personal insurance account.

By having your personal insurance account operate like an investment instrument, you could look forward to an annual Christmas dividend if credit remained in the insurance account. Many people would then calculate their health needs more carefully, acting as if they had a deductible limit, because it would pay them to avoid using the policy.

Insurance companies might find it worthwhile to offer bonuses to people who had not used their insurance for a period of time. It is conceivable that the savings on paperwork in the event that minor claims were not filed might decrease the insurance company's overhead enough to pay for large bonuses to nonclaimants while still making a profit.

Individualizing the insurance system for health care returns cost to the marketplace, as a matter for negotiation between citizen and health-care provider, as it increases the citizen's incentive to stay in good health: People who know that another year without hospitalization might bring them a thousand-dollar bonus may watch their diet and exercise a little more carefully.

This process of shifting the system will pay only small bonuses at first, but as a process it may yield major changes over a ten- or fifteen-year period.

Given a decade to adapt, doctors will be more cost competitive. There will be consumer guides for hospitalization and alternative home-care systems for all but major illness. If we shift the focus of government to health maintenance and to self-care, we will end the upward spiral in health-care costs. In addition, if we eliminate the great bureaucracies, we will see a flowering of new opportunities for citizens to lead longer, healthier lives.

Rethinking Education

The most important step in improving education today is not merit pay, tuition tax credits, or increased funds at any level, but rethinking the education process and system. You would expect educators to be in favor of intellectually re-examining education; in reality, educators tend to be a special interest like any other.

Woodrow Wilson once commented on the political nature of academic faculties when he said that the United States Senate was far easier to deal with as

President of the United States than the Princeton Senate was to deal with as President of Princeton. Academicians tend to take pride in that comment, but I don't think Wilson was referring to the intellectual prowess of the academic world.

The various education associations tend to be pork-barrel organizations, passionate for funding and interested only in issues of immediate concern to their members. In virtually every state, one of the most powerful, narrowly focused, and money-oriented institutions is the state teachers' association and/or union.

Part of the focus on money is a defense mechanism. School teachers have been trapped by three major characteristics of their profession: first, as a physically undemanding profession, education has historically been underpaid; second, as the most manpower-intensive public bureaucracy, the only avenue of additional salary has been politics and lobbying; third, education has remained a cottage industry that is labor-intensive and capital-thin. Lacking the capital intensity and technological orientation of medical doctors as well as the political clout of lawyer-legislators to simply coerce the community at large, teachers have become our lowest paid professionals.

The welfare-state response to this relative poverty is to organize more teachers, to push harder for political muscle, and to force higher prices for the same services—an answer patently wrong for both the country and the teachers.

It is wrong for the country because it locks the education debate into a pork-barrel fight over money rather than an intellectual dialogue about the future of learning, thus focusing the country on the wrong issues.

It is wrong for teachers because, in the long run, their profession must adapt to become profitable. The process of organization and unionization is forcing teachers in the wrong directions. Further, the National Education Association and the American Federation of Teachers are trying to unionize teachers at a time when society is increasingly dissatisfied with the quality of education.

Instead of focusing on specific changes within the current system, perhaps we should recall the founding of the public school systems in Massachusetts in the 1840s and examine why citizens created a universal education system: the goal of that system was to educate a skilled work force for the economy and an enlightened citizenry for self-government. We should rededicate ourselves to ensuring that every citizen has a chance to learn.

Let me suggest eleven steps which would significantly improve the state of learning in America.

1. *Shift the focus of education from teaching to learning.*

We must shift our focus from the teacher to the student. A teacher-focused system overemphasizes the structure of academic disciplines and the bureaucratic lock step of traditional classrooms, a nine-month school year, and annual promotions to higher classes.

There are many strategies for successful learning—a reward system for learning, for instance, could change the speed with which individuals focus on achieving their next objective.

One experiment worth trying would be to offer a $500 bonus for any child who enters the first grade reading at a fourth-grade level. If we combine that with an October "Achievement Sunday" in some local churches at which the bonus winners are recognized, we will have created a social and economic reward system that can have real impact in poor neighborhoods and change the reading patterns of entire communities.

People who suggest that economic rewards would focus society in the wrong direction are kidding themselves: when children learn that basketball players may earn $1,000,000, while studying reading gets you a grade on paper, it's no wonder the summer is spent playing basketball instead of reading. The tragedy is that few players will make the National Basketball Association. If the same amount of energy were spent in learning marketable skills, the whole community would be far richer.

School authorities will talk about "age grouping" as a reason not to start children in higher grades. That is precisely why schools are failing. As long as children are told that *what* they learn is less relevant than how old they are, it should not surprise us that children believe learning is less important than bureaucratic rules. They are being taught that education is a secondary function of the educational system.

If learning was once again the prime focus of public school systems, we would find many students learning very rapidly and leaving the system at a relatively early age. We would find other students dropping out and getting jobs for a few years until they matured. Teachers could focus on designing learning experiences for the mid-level student who sincerely wants to learn but who has not yet found the keys to knowledge on his or her own.

If our system is restructured to reward the willing, dedicated learner, more students will learn how to learn. Then we will be able once again to lead the world in the amount and quality of our national knowledge.

2. *Lifetime learning and the end of assumptions about age*

As our culture and society change with new technology, new medicine, and new government rules, people will have to keep learning until they die. One of the reasons that Democratic scare tactics about Social Security were successful in 1982 was that many of the current generation of senior citizens lack adequate learning skills. Therefore, they are easily frightened by any proposed change.

The challenge to keep learning will be even larger in the future. Our grandchildren will have to plan multiple careers in a working lifetime. We all must continue to learn, because whatever job we have today may not be the same job we must do tomorrow.

In addition, the shift from an industrial to an information society will make it easier as well as more profitable to learn. The development of the home computer, cable television, the telephone system, and the projected public-library network will exponentially increase the knowledge available to average people in their living rooms. Those citizens who keep learning will save money on their income tax, earn more, and have more.

Education systems must adjust to the need for lifetime learning. Bright students will graduate sooner and immature students will be able to drop out and work at menial jobs, knowing that they may go back to school when they mature. Retraining will become an integral part of the unemployment system, and people with current jobs will be encouraged to learn so they will always be prepared if their industry changes.

3. *Focus on discipline and the fundamentals.*

Schools must stop trying to baby-sit, prepare children for life, provide them with ethical guidance, and introduce them to dozens of interesting courses. The more duties we require of schools, the less well they do any of them. Proliferation of objectives leads to decay of performance.

The priorities for schools should be mastery of learning fundamentals and the development of academic discipline.

We don't need to go back to the fundamentals of 1900. Instead, we need to move forward to the Triliteracy which Toffler prescribes for our schools—the traditional basics of reading, writing, and arithmetic alongside the new basics of computer literacy and information management. The truly autonomous citizen of the 21st century must be able to handle all five layers of competence to continue learning at the pace and in the systems which will dominate the information society.

Discipline can be taught partly in the classroom, but it is also the core reason for certain extracurricular activities. Any endeavor which requires practice and determination, such as football or band, teaches children how to persevere in adversity.

If students leave school with the ability to read and write, to understand basic arithmetic, to use a computer, and to *use* the television and radio they deal with every day, then those students will have acquired the core understanding and skills necessary in the future. If, in addition, the student has acquired self-discipline through a series of socializing experiences, then a good citizen has been formed.

4. *From state-focused teaching to society-supported learning.*

Liberalism has weakened the nation's ability to reach educational goals by restricting itself to a bureaucratic approach to accomplishing them. There are a host of learning opportunities available in our society which the education bureaucracy simply ignores: public libraries, newspapers, and educational television, to name just a few.

Further, we need a much closer relationship between the business community and education community if we are to retain any kind of technological ability. Most engineering schools and scientific laboratories simply cannot afford to invest the kind of money required to buy modern equipment. Since, in some fields, there can be entire revolutions in technology every few years, it does not take long for laboratories to become hopelessly out of date, with the result that many of today's college laboratories resemble 1950 more than they do 1984 and our young engineers and scientists, schooled on obsolete equipment, must be thoroughly retrained when they enter the business world and encounter modern equipment.

We should make it easy and profitable for businesses to share their investments in equipment with the academic world. Today, tradition, bureaucracy, and laws discourage joint use of business equipment on a large scale.

Furthermore, present-day education keeps people from contributing their talent to our schools: our school employment rules are so ill considered that a native German with no teaching credits would be barred from teaching German, while a teacher who could not speak German but had all the necessary accreditation would be hired. Society should not, for bureaucratic reasons, bar qualified people from sharing their talents while paying less skilled people to do jobs for which they are unqualified.

Finally, there are many decent citizens who would be glad to help if only we could organize their participation. For instance, we could build a senior citizens' volunteer corps to tutor students. There are many grandparents with key skills and disciplines. Some of them would share their talents for two hours a week tutoring students.

In sum, by gathering our society's resources to encourage education, much more learning could occur with little increase in cost. It is time to try.

5. *Re-establish apprenticeship systems.*

One of the problems which keeps the poor poverty-stricken has been the decline of apprenticeship programs. Good work habits and invisible assets such as networks of contacts can be developed in an apprenticeship program more easily than in school.

Many skills and trades are learned better through apprenticeship than in academia; many disciplines are rendered lifeless and inaccurate by an academic approach: cooking and politics, for example, are best taught by a practitioner.

There are fine young people who simply do not fit well into school but would love to learn a trade. An opportunity society would develop apprenticeship programs as an alternative to high school, offering tax advantages and subsidies to both the apprenticer and the community.

Apprenticeship should be combined with a strong learner-focused education program for all ages using the public library, home computers, television, and radio programs. That way apprenticeship programs would not cut adolescents off from other avenues of education.

At a time when one out of five Americans is functionally illiterate, we must try something new.

6. *Using mass media and computers for learning*

We have grossly neglected the potential educational resources of new technologies.

Today, learning can be had with minimal inconvenience by people in hospital and nursing-home beds, by mothers watching their children, by underemployed people who want to learn at night, and by people who wish to expand their knowledge for fun. The potential linkage between the home computer, telephone, cable television, and the local newspaper is enormous. For a modest sum, we could build a software package that would make a large library available to home computers by telephone hookup.

For instance, people with particular illnesses might wish to learn from computer and videotape about the problems of their illnesses, the potential complications, and the correct way to take care of themselves.

The government could use public television and direct mail to offer people in areas of high unemployment access to knowledge about new jobs and new skills.

There is a great world of educational opportunity just beyond our reach. At a minimum, we should begin developing experiments in community-oriented technologies.

7. *Challenge higher education.*

Today, students study political science instead of either politics or government. They study sociology instead of societies. The academic world has become so self-important that it thinks an introduction to the history of *its* ideas is the same as an introduction to the history *of* ideas.

While I have philosophical doubts about reductionism as the best way to understand the world, at least in the reductionist sciences students learn a rigorous method of viewing the world. A physics student or a biology student may see only a slice of reality, but the emphasis is on reality.

Reality does not interfere with a typical course in education or social studies ("social science" is an arrogant misnomer). In these areas, students spend their time learning about the structure of knowledge rather than the structure of the world. In fact, the structure of knowledge *as taught* is merely the structure of the discipline being taught. Students are thereby further distanced from reality.

The current model of intellectual disciplines was an invention of the late 19th century and is supported by the political power of the academic bureaucracies. It would be healthy to debate alternative ways of organizing learning.

The 1870s Harvard innovation of the modern undergraduate curriculum is ready for an overhaul. The 1880s German invention of the modern Ph.D. program is equally ready for a challenge. The time has come for some college or university to seek alternative methods of encouraging students to learn about their world and about learning.

8. *An intellectually open academe*

One of the surprising things about the modern academic world is that it tends to be as ideologically monolithic as a Republican country club. We are still awaiting someone to write the *Main Street* and *Babbitt* of academe. Babbit's grandson, now a Ph.D. in Sociology, earns a pittance, lives in genteel poverty, drives an old Volvo, and attends a Thursday afternoon wine-and-cheese party at a faculty member's home. He feels morally superior to his high school classmates who entered business, medicine, and law: they earn more than he does, but he *knows* more.

The arrogance and cynicism of the modern academic is the most disheartening aspect of the current system. The left-wing ideological biases of the academic community are now as much a barrier to truth as were the right-wing biases of the business community in the 1920s. The search for truth requires recognition that there are alternative truths. On many campuses, the only truth is left-wing, mandated to teach contempt for middle America's traditional values.

The debate required is not one of Conservative politics versus Liberal politics. The problem is that we need an intellectual struggle in the academic community, not a political struggle.

James Q. Wilson, in *Thinking about Crime,* made a solid case that Liberal biases made it impossible for sociologists and criminologists to deal with the data they developed. Sociologists tended to start with their personal biases as an opening chapter, develop their data in the core of the book, then revert to their own biases in the concluding chapter. There was often either no relationship between the data and the recommendations or a direct contradiction between the data and the ideologically derived conclusions.

Here are four specific questions for a true intellectual debate:

1. Which identifiable ethnic minorities in modern America have risen faster than the Black community? Asiatic refugees, Chinese and Japanese Americans (especially those who were uprooted in World War II) and West Indian Blacks are ethnic minorities which seem to have risen faster than the Black American community. What lessons for public policy can we learn from these communities?

2. What lessons did the Vietnam War teach us about the American use of power? Increasingly, most of the serious writing about military history in Southeast Asia is being done by professional military theorists who are trying to formulate lessons for the future. The North Vietnamese have admitted that they thought they had lost the Tet Offensive, that they consciously used the American news media and the American Left, and that the war was planned in Hanoi rather than in the South. Nevertheless, the academic world still behaves as though Tom Hayden were right. A revisionist historical debate using new data is badly needed.

3. What is the impact of the left-wing bias of television? The intellectual community would react fiercely if prime-time entertainment were as biased toward Jerry Falwell's values as it is currently biased against. Present research into the electronic media revolution is thoroughly inadequate.

4. How can we study economics in a way which

takes into account interest-rate expectations and the behavior of entrepreneurs? The current model of economics is too mathematical, with an orientation toward the quantifiable, the governmentally generated, and the statistically available. The economy is more psychological and cultural than current econometric models admit, with the result that much of our economic advice is closer to reading chicken bones than to serious science. Econometrics do not explain the growth rates of Hong Kong, Taiwan, Singapore, and South Korea. Orthodox economists were sure the West German economic experiment in the late 1940s would fail; it led to a boom instead. Present economics, as Peter Drucker suggests, is only useful for linear projection of what you already know. All the big questions for society involve what you don't know.

These are just a small sample of the kind of questions which current biases prevent the academic community from studying. It is time to open up the intellectual world to intellectual exploration.

9. *Learning for everyone*

One of the tragedies of the modern academic world has been its devaluation of liberal arts education. People should be encouraged to read history so that they know something about the world in which they live and fiction to learn more about themselves and about people in general. To the extent that liberal arts degrees have become vocational, such degrees have been trivialized. We are all the poorer for it.

Not only should a liberal arts education last a lifetime, it should also take a lifetime. Benjamin Franklin established the American Philosophical Society, in part, so that people would have a place to go and talk about ideas. We should challenge adults in every walk of life to take up the pursuit of ideas once again. A generation from now, every local country club and neighborhood recreation center should have an idea center as well as a tennis center.

10. *Knowledge from action research*

Lynn White notes that medieval agriculture progressed fifteen hundred years beyond medieval academic understanding *of* agriculture. Academicians were still reading Aristotle to learn about horses when there were modern horses at the hitching posts outside their classrooms. Since medieval agriculture was making great advances in plowing and in the use of the horse, it was the impractical man who listened to the academic.

We face a similar problem today. Far too much of the academic world is spent on self-analysis rather than in studying the world at large. Far more advances are being made in the world by practitioners than are being developed by academics.

We must return to a reality-based approach to knowledge, especially in business, government, politics, society, and economics. Real world research must take precedence over theoretical papers. The academic world of today has traded relevence for introspection. What you know and whether or not you are right should become as important as where you got a degree. Memoirs by first-grade teachers should be as academically important as analyses of questionaires of first grade-teachers by intellectuals who have never taught first grade.

The time has come to leave Aristotle and go back to the horse for a while. Action research, leading to reality-based publishing, is the first step.

11. *Change and traditional values in learning*

Peters and Waterman, in *In Search of Excellence,* show that excellent managers insist vehemently on core values but then are very loose and flexible on everything else.

The core values of public education should include a focus on the student rather than the teacher, use of all the resources of the community, lifetime availability for everyone, and provision of a thorough grounding in Triliteracy. Beyond these fundamental commitments, we should encourage public learning to be flexible and innovative.

The academic community is now locked into bureaucratic models from first grade through Ph.D. programs. Academe has become an increasingly self-centered bureaucracy, tolerating less and less interaction with the world around it.

No single reform will adequately change the current system. In fact, an effort to develop a national answer would prove self-defeating. What we need is a multitude of experiments involving computers, television, apprenticeships, action research, the business community's equipment and personnel, and especially the individual citizen, who has competence but no credentials.

Data prove conclusively that we are the first generation in American history to educate our children less well than we were educated. Of the major Western nations, we score close to the bottom in math and science. We know what we are doing is not working.

The structural reforms I suggest will lead to many failures, but only by risking failures can we advance. We will only achieve excellence if we encourage innovation. Our children and our nation deserve no less.

From Welfare to Workfare

We must compassionately reconsider poverty, unemployment, and the culture of being poor. Our present, and inadequate, solutions stem from the New Deal, an era with different mores and the pressures of the Great Depression. These solutions are no longer useful and must be reformed.

The welfare system has a number of unintended side effects, second- and third-order consequences which went unanticipated by its original planners. For example:

Efforts which originally were meant to ensure that only the truly needy received assistance have since become rules which drive men away from their families.

Efforts to ensure that only those without resources were helped had the unintentional effect of penalizing individuals who had saved only enough to bury themselves.

Efforts to ensure that only the destitute aged were cared for by the government forced persons without incomes to cede what little they owned to their children in order to be permitted to move into nursing homes at government expense.

Efforts to ensure that young girls who were pregnant received adequate care became incentives for teenage pregnancy.

Efforts to protect workers from temporary unemployment became linked with union-negotiated programs which encouraged laid-off workers to remain tied to stagnating industries.

Efforts to provide adequate unemployment compensation became incentives to avoid work.

Efforts to hire the best professionals to plan care for the poor turned into disasters like the Pruett-Igoe housing project in St. Louis, created by architects who neither had lived in public housing nor understood the problems of a poor person in a large project.

Efforts to relieve the poverty of truly disadvantaged Americans devolved into across-the-board programs: for instance, one project under President Carter gave "energy assistance checks" to residents of public housing projects whose energy bills were already paid by the government.

Let me suggest just a few guidelines around which we might build an opportunity-society welfare system:

1. Individuals, families, and poor communities should receive cash and credit card vouchers directly, to increase their choices and give them power over their own lives.

2. Systems should be designed on the assumption that people will yield to temptation and that cheating *will* occur if the only deterrent is bureaucratic enforcement. Therefore, whenever possible, the incentives engineered into the system should enforce themselves.

3. These services going to the very neediest should be seen as state charity and should be generous rather than stingy. Our current system crimps those at the

bottom of the welfare ladder. No one must fall beneath a certain level of poverty, even if we must give away food and money to prevent that from occurring.

4. Those who are elderly, or have disabilities or handicaps, should be helped cheerfully and without expectation of return. Those who have disabilities or handicaps should be retrained and investments made to allow them to lead fully productive lives. Those who are elderly should be cared for on the grounds that it does us honor to be able to help our parents and grandparents. It is a sick society which creates for itself a future of fear and insecurity.

5. Those who are able-bodied and under the age of retirement should work or study if they wish to receive aid—including unemployment compensation. Among other things, this principle will require the establishment of day care centers where some welfare mothers care for the neighborhood children while others go out to work.

6. Minor children should be eligible for welfare only through their families or guardians—this one step would end government's unintentional encouragement of teenage pregnancies.

7. People should pay something for every health service, even if only fifty cents. If people are too poor to pay fifty cents, then, under an earlier principle, we should raise their income so they could afford to pay for needed services. Free service is an invitation to abuse the service.

8. The tax and welfare laws should be changed so that those trying to work their way out of the welfare system gain more than they lose at each step. Taxes, child care, and the cost of going to work should be set against gross pay in calculating welfare eligibility. Reinstating the tax deduction for dependents at the constant-dollar value it had under Harry Truman (equivalent to $4800 in 1982 dollars) would automatically eliminate taxes for most of the working poor with children. Eligibility for health care, housing, and other programs should be changed so that there is a positive bias in favor of workers. The current system encourages an underground econ-

omy in which people take unreported cash jobs in order to retain their welfare and unemployment benefits, the surrender of which cannot be justified by the job's pay. We should make it legal for people to do what they are going to do anyway.

9. We should help poor people take real control over their lives. Every step toward empowering people to control their immediate environment is a step toward increasing their self-respect and their desire to rise to a better status. For example, most public housing projects should become condominiums which poor people can buy through sweat equity and which their housing association would then run. This would promote self-respect, protection of property, and a sense of community, and it would involve people in government.

10. Whenever possible, resources should go directly to the poor, who should be expected to guide their own lives. If we abolished every anti-poverty agency in America and turned the same amount of money over to the poor in individual credit cards, not only would the recipients have a much better chance of improving their lives, but also society would save the enormous overhead costs, and the poor would be treated with the dignity they deserve.

11. We should consider a nutrition program which made the most basic foodstuffs—flour, cheese, rice, red beans, and a dozen similar items—available free or at nominal cost in grocery stores to anyone who wanted them. This would cost far less than the current twelve billion dollars annually for food stamps. There would be no excuse for hunger in America; there would be no food-stamp fraud; and there would be no popular fury at seeing luxury foods paid for, on occasion, by food stamps.

12. We should involve public television and radio in bringing useful information to the poor, such as how to install cheap insulation, how to buy and cook good meals at low prices, how to take care of personal health, and how to establish credit.

13. Individual actions should relate to individual responsibilities. For instance, we should reshape child

support laws so that parents are forced to support their children. When one parent—almost invariably the father—abandons the child, we should use the Federal government to track him across jurisdictions. We should require the non-supporting spouse to pay any court and legal fees, any late-payment costs the ex-spouse has incurred, and any interest penalty as well. The same principle should apply to illegitimate children. In short, parents should be tied to their own children.

14. We should develop more part-time work programs in lieu of most of our current welfare costs: people receiving aid can be required to serve as day-care workers, members of a neighborhood watch, aides in nursing homes, or in similar positions which increase the quality of life for the community, reestablish a relationship between work and economic reward, and give people a chance to acquire skills.

15. Most punishment of non-violent crimes should become an act of restitution; simple hard work, such as helping the neighbor you vandalized, will improve the quality of the neighbor's life, teach you that crime is expensive, and keep you out of prison. Prisons should be holding tanks for violent and repeat offenders.

Our goal must be to help the helpless and the elderly so that they can lead full lives and to empower the poor to control their lives and rise from poverty.

If we design the right programs now, then no embers of the next generation will be able to say they live in poverty because they have no choice.

Military Reform: Tackling the Largest Bureaucracy

While the Department of Health and Human Services may disburse more money in the form of transfer payments, there is no other system in the Federal government remotely comparable to the Defense Department in size and complexity.

Everything conservatives have said about bureaucracies in general applies in particular to the Pentagon: it is well protected by its allies on Capitol Hill and in the military-industrial complex; it is hide-bound by traditions, some of them going back nearly two hundred years; and it is a hive of internal political squabbles. The perceived enemies of Pentagon planners appear to be other services and factions in the Pentagon rather than the terrorists and totalitarians who threaten the rest of American society.

Robert MacNamara imposed on the Pentagon the same Harvard Business School management techniques which almost destroyed the American automobile industry. Further, Congress over the last thirty years has treated the military to pork-barrel politics and mindless micromanagement which combine to make it impossible for the Pentagon to run efficiently.

Congress should pass a multi-year budget and let the Pentagon try to manage itself for the period of that budget. The experiment would certainly be no more costly and wasteful than the current circus of overmanagement, with Congress as ringmaster.

There are some weaknesses in the current military system that need to be dealt with directly. The most serious is the under-use of the national guard and the reserves. A regular-duty soldier is nine times as expensive as a reservist or a guardsman. Regular soldiers are paid every day, while the reserves and guard are paid only when they train. Further, the regulars retire after twenty years on half-pay pensions. Our military today is too small to meet all its potential commitments, and too expensive besides.

We must reform Pentagon purchasing through testing in fly-offs and shoot-offs before buying. There must be more delegation to the field commanders and developmental commanders (such as the 9th Test Bed Division concept). There must be more procurement of off-the-shelf items. There must be more multi-year procurement.

The time is ripe for a thorough overhaul of our military management system. There is no excuse for focusing on the smaller bureaucracies while ignoring this one.

Deregulating the Federal Bureaucracy

Most Federal civil servants want to do a good job and are as totally frustrated by irrational, obsolete bureaucratic systems as the people ill-served by them. Sooner or later we must start trusting our civil servants and trying to work with them.

A free people may indicate public policy through their elected officials, but they can only implement those policies through a professional staff. In the information age, even a lean Federal government is going to be an enormous structure to manage.

We have the capacity to revamp the Federal government into an exciting, dynamic place to work. If we combine technology with a system of merit pay, idea bonuses, individual retirement accounts for those who desire them, and other steps toward a more flexible and desirable style of work, then I think that, in the 21st century, we can once again be proud of the quality of our civil service.

In each area of government there are great opportunities for an age of improvement. The time to start making those improvements is now.

CHAPTER 7: WHY BALANCING THE BUDGET IS VITAL

For fifty years, Conservatives have made balancing the budget a key part of their political platform. They have preached that red ink and big deficits are bad for America. They have reminded the public that we can't run our family budgets with a constant deficit and neither can the country. They have been right.

Now that Republicans are in power in the White House and share power in the Senate, they are finding it remarkably difficult to live up to fifty years of preaching. Ronald Reagan, who spent twenty years attacking the big-spending Federal government, is now presiding over the fastest increase in the Federal debt in American history. Why is there such a contradiction between intentions and reality?

The basic reason that the Federal deficit is uncontrollable is that Federal spending is uncontrollable. Washington is a city dedicated to picking the pockets of average taxpayers: every special interest group in Washington gains power and influence by raising spending.

Washington is an increasingly imperial city which considers it the duty of the rest of the country to pay tribute so that Washington can raise spending. This attitude is shared by lobbyists at cocktail parties, reporters in the press gallery, committee staffers, and bureaucracies of the executive branch. Special interest groups work closely with Congressional committees which provide money for their interests, with special-interest media which report on appropriations, and with the executive-branch agency which will then disburse the money.

The result is an iron triangle of special interest: anything which threatens the increased expenditure of public funds is, by the interest group's definition, bad.

Since the various partisan interests mix at one another's parties, they share a general mindset which begins with the premise that "Nothing can be done about spending." Given that presumption, the natural corollary is that "We will simply have to raise taxes."

If this were a Liberal country being presided over by a Liberal capital, then that equation would be correct. However, the fact is that Americans are a conservative people who are increasingly angry that their taxes have gone up without curing the deficit problem.

The average American family's income taxes were 6 percent of gross income in 1958 ($303 of a $5,087 median income) and had risen to 9.2 percent in 1983 ($2,218 of a $24,100 median income). In other words, income tax dollars increased sevenfold, in part because of inflation and because the percentage of family income taken by income taxes has gone up by half. In addition, Social Security taxes have increased even faster, to the point that they now constitute the largest tax burden for more than half of all American workers.

However, the problem of our government profligacy has increased so rapidly that even this larger share of income is totally inadequate.

In 1958, a balanced budget would have required taking 6.5 percent of the median income in income tax. That was only 8.3 percent more than the actual tax burden. In 1983, however, a budget balanced through income tax increases would have taken 15.9 percent of the median income in taxes, a 72.8 percent increase.

The case against tax increases is even stronger once you consider the impact of inflation on real family income. Most of the increase in family income during the inflation of the 1970s was illusory, but that still pushed taxpayers into higher tax brackets.

One of the reasons for the tax revolt of the 1970s was that working families, formerly below the level of serious income taxation, suddenly found themselves in brackets which cost them real money. Families often found themselves getting pay raises too small to keep up with inflation but just enough to push them into higher tax brackets. The result was a double drop in real take-home pay.

Led by men like Ronald Reagan and Jack Kemp, Proposition 13 in California, Proposition 2½ in Massachusetts, and similar movements—including the Kemp-Roth three-year Federal tax-cut proposal—became popular because people found themselves running harder and harder, but farther and farther behind.

Consider a typical family of four that was earning $16,000 in 1971 and whose income kept exact pace with inflation. The family's real income in 1981 was the same $16,000 in 1971 dollars; consequently, the family was in real trouble.

Without having increased its real earnings one dime, inflation had pushed this family from paying 19.1 percent of its income up to a bracket where it was paying 29.1 percent of its income in Federal, state, and local taxes.

In order to accomodate a 56 percent increase in the government's share of its income, the typical household had to shrink its spending on clothing by 40.7 percent, its spending on housing by 11.6 percent, and its spending on food by 3.5 percent.

In other words, if you felt that you were trying to live on less and less in the 1970s, you were probably right: inflation was eating away at the real value of your money and the government was competing for whatever was left.

Since people have come to feel that bureaucracy is inefficient and that the welfare state tends to take their money for causes that the average worker does not endorse, there has been a natural resistance to tax increases. Most people simply do not believe they are getting their money's worth for the taxes they already pay.

One recent poll showed that, when asked how to balance the budget, 60 percent of those polled said, "Hold down spending," and only 7 percent said, "Raise taxes." Sophisticates in Washington claim this question is naive because people are always *for* spending cuts in general but *against* cuts in their particular interests. Washington professionals "know" that some provincials may still believe government can be mastered and spending controlled, but that this is not possible in modern Washington. Washingtonians concur that only by raising taxes can we offer any realistic hope of shrinking the deficit.

Frankly, the nay-sayers are correct, under present circumstances, because it is impossible to control spending in the welfare state. Every political pressure and cultural bias is in favor of more, rather than less, spending. The dominant political coalitions are held together by the promise of more Federal money.

The only way to bring Federal spending under control is to have a political revolution in Washington. Just as it would have been impossible to run a Liberal welfare state administration under Calvin Coolidge, it is equally impossible to run an opportunity society under the present House Democratic leadership and the current staffs of the Congress and the executive branch.

The choice, therefore, becomes one of huge deficits with their destructive effects on interest rates, inflation, and the economy; or hugely increased taxes to cover the welfare state spending; or a political revolution at the ballot box which permits a real transfer of power from the lobbyists, congressional committees, and executive bureaucrats—back to the people who pay the taxes.

My suspicion is (and virtually every poll corroborates this hunch) that most Americans would applaud an overthrow of the current wasteful, arrogant power structure which is the version of Washington they see in the news media.

When I visit Georgia high schools with many poor students (both white and black), I ask how many have held a job. Lots of hands go up. I ask how many noticed the difference between what they thought they were earning and what they were allowed to take home in their first paycheck. The same hands go up; people start laughing. I then ask how many of them want me to raise their taxes and further shrink their take-home pay in order to buy more food stamps and public housing. No hands go up.

The only hands I have ever seen raised in favor of increasing taxes were those of a college professor and of a wealthy Republican.

The answer to the deficit is not higher taxes, but controlling spending. While some people argue that deficits don't matter, the historic reality is that they are simply wrong.

Some Reagan Administration appointees give speeches explaining that big deficits don't really matter, a line of reasoning grounded in misconceptions about the special circumstances of other countries. For instance, Japan, a nation which saves 19.1 percent of its annual income, can afford to run a deficit for a long time because the Japanese savers use the deficit as a means of earning interest: the Japanese can afford government deficits because of their personal and family surpluses.

In the long run, in the U.S. with a more limited savings rate, large deficits inevitably sop up money which otherwise might be invested in new factories and jobs. Big deficits distort all other economic activity: the carrying costs of debt become a major consideration in public life; as the interest payments increase, government finds itself crowded out of alternative spending by the need to pay interest on its own bonds; ultimately, the government must either borrow so much money that interest rates skyrocket or inflate our currency and cheat people by paying them back with dollars less valuable than the ones the government originally borrowed.

The great inflation of the 1970s set a worldwide, four-hundred-year record. It was a perfect example of stealing from people who save: if you loaned the government money in 1967, you had to receive three times as much in 1983 just to break even. Actually, with taxes, you would have had to have earned even more to break even. The government robbed you by cutting the value of the paper money it was giving you and then taxing what was left.

When people learn that government robs them while they work or save, it is little wonder that an underground economy flourishes. People cheat on their taxes, shift from saving to speculation, and are willing to go bankrupt or undertake any other kind of behavior which will give them a break because they know their government will not give them a break.

The Reagan Administration has been afraid to balance the budget because it has been afraid to break the back of the Washington establishment. We will never be able to balance the budget—or even significantly shrink the deficit—unless we either give in to the Liberals by raising taxes or break the power of the Liberals.

In 1982, Senate Republicans and sophisticated insiders in the Reagan Administration formed "the gang of seventeen" intent on averting the budget crisis through a bipartisan plan.

As a prerequisite for their support, Liberals demanded an overhaul of military programs and an increase in taxes before they would even consider the domestic-spending side of the ledger.

Reagan supported increased military spending and lower taxes, so there seemed little room for negotiation. However, a grand bipartisan compromise was finally announced in July, 1982: taxes would go up by billions, but there would be three dollars in spending cuts for every dollar in tax increase. A number of us were convinced that the Administration was being hoodwinked. Reagan, however, was so convinced of the rightness of the compromise that he went on television and fought the conservatives of his own party—the biggest single mistake of his term.

The 1982 tax reform bill, aimed directly at conservative voters, included measures to withhold interest on savings which, in 1983, led to the largest Populist uprising of our time.

And, despite all their promises to cut spending $3 for every $1 in tax increases, the Liberals engineered a dollar in spending increase for every dollar in tax increase.

Thus, the effort to reduce the deficit raised both taxes and spending; the imperial Washington spending machine was richer, more powerful, and more arrogant than ever; the American worker was both poorer and more cynical. It was a rout comparable to Bull Run, a triumph of the welfare-state mentality.

The reality is that welfare-state Liberals will not voluntarily shrink their own power base, anger their own special interests, or give up their ideological beliefs.

We must replace the welfare state with an opportunity society, and thus replace a liberal budget with a conservative budget.

It is impossible to design a cheap welfare-state budget. As long as the welfare state is the dominant ideological force in American politics and government, it is impossible for Republican budget-balancing to succeed.

The David Stockman budgets have been strategic failures because they did not shift the terms of the debate. As long as the central issue is "How much are you getting from government?" then it is impossible for a conservative administration to develop a successful strategy.

The record of Republican administrations since 1952 has been one of inheriting Liberal economic messes, causing enough pain to arouse all the Liberal allies, irritating conservative allies with minor tax increases and compromises with Liberalism—and failing to create a new dream around which to rally the American people.

Eisenhower, Nixon, Ford, and now Reagan have all fought to tighten government purse strings without changing America. Nixon made a daring effort to create a center-right coalition by purchasing the support of the construction unions with the Davis-Bacon Act, supporting a variety of domestic programs, and seeking to build a cautious-but-still-positive welfare state. The result for Nixon in 1972 was a landslide, but Republicans lost three Senate seats and gained only twelve seats in the House. Nixon gained personal acceptance, but it proved impossible to build a conservative victory around a modified welfare state.

Eisenhower, after 1958, and Ford, during his short presidency, were reduced to governing by veto. While both made a pretty good case that they were protecting the people from higher taxes and inflation, the Republican Party lost the following Presidential election in both cases. A veto strategy can only be a defensive strategy unless, as Harry Truman proved in 1948, it is deliberately used to set the scene for a positive vision of America's future.

Despite the belief of some Republican loyalists, "stopping government spending" and "holding the line on the deficit" are not battle cries around which to rally a nation. The genius of the Kemp-Roth bill was that it began to turn the debate from "How much do you want to spend on each interest group?" to "Do you really want to raise the taxes of these hard-working Americans and thus discourage growth in jobs and opportunity?"—a change of focus which is enormously to the advantage of the conservative movement.

Our vision of America must be one which proffers tax cuts, lower interest rates, and honest money while we hammer our opponents for favoring higher taxes, higher interest rates, and dishonest money.

An opportunity-society coalition must be built around honest money, paychecks, and savings-interest payments. We want people to earn more real money. Experienced savers would accept lower interest rates if the interest came in honest money without additional taxes: they could keep the money and the money would be real. All citizens want to increase their take-home pay in real dollars.

A working-saving coalition would unite the majority of senior citizens with savings—those worst-hit by inflation—and younger workers who would save if they had enough take-home pay. Just as the McKinley majority was built around a full dinner pail for the working American of 1896, so a new majority should be rallied around the full take-home paycheck and the full savings dividend. Liberals would be left with people who intended neither to earn nor to save.

Seen in this context, every effort to control waste in government and runaway spending would be a step toward more take-home pay, sound money, and a better break for savers. This would encourage people to work, save, invest, develop new products and technology, and create jobs.

This new platform could emphasize that people had been put to work because of lower interest rates and higher job investment. Conservatives could show that they had created the most effective jobs program in history and could ask why Liberals wanted to cripple a system that was working.

Most farmers would be intrigued by a campaign to lower their interest rates. It has been estimated that each percent in interest rates is equal to 1/10 of total farm income because the American farm community is now so heavily in debt. A sharp drop in government borrowing, followed by lower interest rates, would lead to more prosperity on the farm than all the existing government farm-aid programs combined.

Similarly, most auto workers would rather build cars than draw extended unemployment benefits. If they believed that interest rates would plummet following curbed government spending, they might well vote to restructure everything—including unemployment—in an effort to create new jobs in the auto industry.

Briefly, in 1981, Reagan had put together a coalition of American workers and savers. Lobbyists were coming into my office, asking me to take away benefits from their vested interests. There was a general feeling that all Americans must tighten their belts to get us out of the economic crisis created by the previous administration. There was a feeling that Reagan was doing the right thing even if it hurt a little bit.

Democrats counterattacked brilliantly with a campaign of fear and demagoguery the like of which has not been seen since Joseph McCarthy. In a totally irresponsible and inaccurate manner, they began characterizing any effort to curb spending as a general campaign to hurt the poor of America. Democrats discovered that they could manipulate Blacks with the specter of a return to segregation and senior citizens with the abolition of Social Security. Those two steps greatly improved Liberal fortunes.

The Reagan Administration's failure to explain its program gave its Liberal opponents ample room to attack. What began as an effort to get the Carter Administration's outrageous increases in spending under control degenerated into an orgy of news-media assaults on cheapness, hardheartedness, and incompetence. The Reagan Administration ended up looking unfair, callous, and indifferent to the needs of average and poor Americans.

All this was avoidable, but an opportunity society will have to find a way to explain its goals if ever we are to balance the budget.

There are some clear historical examples of successful politicians who represented working-saving citizens' coalitions. In Britain, the administrations of Walpole, in the 18th century, and Peel/Gladstone, in the 19th century, both took over governments in serious economic difficulties and with enormous debts; both instilled in those governments a sense of legitimate frugality and their publics responded as their economies grew and pocketbooks were filled with money their governments did not take.

Walpole, Peel, and Gladstone built public trust and developed an approach to spending which spread a sense of caution throughout their governments. All three leaders were careful to avoid overextending their governments in either foreign or domestic policy; all three required that projects demonstrate substantial potential payoffs before they were undertaken; none believed in throwing away the public's money.

One of the problems which the Reagan Administration must confront stems from the fact that a fiscally prudent government must overhaul and reform the

Pentagon: it is impossible to urge frugality on the rest of the American government while ignoring the profligacy which is the hallmark of modern Pentagon mismanagement.

The issue of waste is one of the major reasons a balanced budget must be a serious goal of our government: if a balanced budget is our target, then resources matter; we must set priorities, and be careful with money. The minute we give up on balancing the budget, we tell special-interest groups and bureaucrats that the only limit on their access to Federal money is the amount of political power they can bring to bear. If there is no balanced budget to impose discipline, then every interest group will fight for everything it can get, and every bureaucracy will feel relaxed about wasteful spending.

If we really want to cut out waste and abuse in government, we have to set absolute standards so that no group feels its extra effort will be wiped out by the waste and extravagance of some other group.

One major step toward a balanced budget would be to give the President a line-item veto. With such veto power, the President could eliminate specific budget items. Forty-two of the fifty state governors already have this power to control spending.

Congress now mixes appropriations for unnecessary programs with appropriations for critically needed ones to blackmail the President into accepting the extravagant in order to get the essential.

For example, Congress usually packages unnecessary spending with Social Security spending. Then, for three days, we see news stories about how Social Security checks will stop and people will have no food unless the President signs the bill.

Of course, the way we always solve such a crisis is by spending everything Congress wants to spend. Politically, the President does not dare veto a bill that includes Social Security spending.

Congress has proven incapable of disciplining itself and balancing the budget on its own, but, with a line-item veto, the President can trim annual budgetary excesses and bring us closer to a balanced budget.

Another major step toward national solvency would be a constitutional amendment providing for a balanced budget except in times of war or national emergency.

We need both these constitutional changes in order to regain a sense of balance in Washington: realistically, only these amendments can break the "iron triangles" of special interest.

The challenge we face is difficult, but it is time that the baby-boom generation grew up politically as well as chronologically. There are no easy answers to the mess into which twenty years of excessive welfare-state spending, bureaucracy, and subsidization have gotten us. It will take a sustained and concerted effort to create the world we want to give our grandchildren.

We know, however, that if we proceed along the present track, we will bankrupt our country. Personally, I do not want to explain to my grandchildren why their taxes are enormous, their industries obsolete, their money worthless, interest rates sky high, their bureaucracy arrogant and inefficient, and their prospects dim. Furthermore, I don't want to dislocate my grandchildren's lives that badly and then have to turn to them for my retirement and medical care. If we err grievously now, I suspect that our generation may have a rude awakening when we get ready to retire and our grandchildren tell us to fend for ourselves.

To forestall doubters, let me share a few examples of just how much danger there is in the deficit:

By 1989, interest on the national debt will have reached 178 billion dollars a year, according to the Congressional Budget Office. That means 3.3 percent of all economic activity (the GNP or Gross National Product) of the country will be earmarked for interest on the national debt—only .3 percent less than all the non-defense discretionary spending projected for the entire Federal government in that year.

Interest payments on the national debt will amount, in 1989, to approximately $712 for every American, which means that an average family of four will pay $2,824 in 1989 taxes just for that interest.

To better comprehend the scale of the crisis, consider that the government already spends more on interest than it does on the entire U.S. Army. By 1989, the probable cost of the army, marines, and navy together will be less than that year's interest payments on the debt: interest will be twenty billion dollars more than the entire cost of Medicare and Medicaid.

By 1989, the entire "human services portion of the welfare state (except for Social Security and health care) will be thirty billion dollars less than debt interest payments.

Not only will interest payments in 1989 by high, they will be growing, since the projected deficit for that year is another $280 billion. In other words, every dollar we are spending on interest is a dollar we are borrowing to pay that interest.

In fact, there is not a single year in the 1980s when the deficit is smaller than the debt service. Every year in this decade, we will be borrowing money to pay the interest on money we have already borrowed. By the end of the decade, we will be borrowing money to pay interest on the interest we have failed to pay on the debt we owe. Compound interest is beginning to devour the economy.

Any citizen, family, or business which finds itself so debt-ridden that it has to borrow just to pay the interest on its notes is in deep trouble. Any banker would warn a customer that this kind of behavior leads to disaster. Countries are bigger and take longer to collapse than do individual households, but sooner or later the same rules of economic reality apply.

If the United States remains foolish long enough, sooner or later it will go bankrupt. Countries go bankrupt by repudiating their debts, cheapening their currency, or taxing their citizens into rebellion.

Ironically, it was this issue which aroused the Founding Fathers, complacent at having defeated the British, to realize that they had a new set of problems. Shays' rebellion in Massachusetts was an effort by debtors to repudiate their debts and to force government to inflate the currency.

Jefferson, Washington, John Adams, Benjamin Franklin, and their like were horrified by the prospect of debt repudiation. They believed that a free country could survive only by being an honest country, and that included an honest currency. They also believed that freedom had to be based on hard-working people who were willing to save and invest in their nation.

It was the threat of economic decay, inflation, and repudiation of debt which brought these men together in Philadelphia to write a constitution and found the United States of America.

The time has come to rethink our government and to rebuild our approach to solving society's problems in the tradition of the Constitutional Convention.

CHAPTER 8: CREATING JOBS, HONEST MONEY, AND OTHER ECONOMIC DREAMS

One of the most frustrating failures of the welfare state has been the gap between its economic promises and its economic performance.

The Left has gained political power throughout the Western world by promising to improve economic conditions through a formula of centralized bureaucratic planning and redistribution of wealth: rational, technocratic civil servants would replace wealthy entrepreneurs as the guiding force in the economy; government would take money away from wealthy entrepreneurs and give it to the poor in return for their votes.

This has, on almost every occasion, led to the same discouraging result: technocratic civil servants, vulnerable to political pressures, lacking the skill and incentive necessary to invent the future, fail to perform the promised miracles. The keys to economic growth are dynamic and require the kind of intuitive sensitivity more typical of entrepreneurs than of bureaucrats; socialized economies inevitably stagnate as unemployment increases, ability to compete with foreign countries decreases, dissatisfaction with government grows.

In order to hide the failures of its centralized planners, left-wing politicians then give away more goods and services than the country can afford. At first, they pay for these political bribes with borrowed money, but that drives up interest rates; then, in desperation, the government starts printing paper money and inflation sets in.

Such an economy, weakened by bad central planning, is further crippled by high interest rates which discourage investment in new factories and new jobs. Inflation discourages savings and hard work in favor of speculation and borrowing. In addition, the hostility of bureaucrats and politicans toward entrepreneurs leads to a flight of talent from the country.

Faced with this scenario for economic disaster, the solutions offered by the Right have had three problems.

First, conservative administrations have consistently focused on undoing Liberal mistakes rather than on creating conservative solutions. In Irving Kristol's phrase, they have become bookkeepers of the welfare-state budget, which is hardly a desirable image. The Right has failed to build a positive coalition of forces which would benefit from a more conservative economic policy. Indeed, at times in pursuit of their first job as balancers of welfare-state budgets, right-wing politicians have gone out of their way to offend their own potential allies (e.g., withholding on interest on savings).

By far the most important failure of conservative regimes since World War II has been their inability to reshape the public understanding of how the economy works. Margaret Thatcher, the conservative leader of our time who best understands the importance of educating as well as administering a country, asserts that you can't just win the vote, you also have to win the argument. Indeed, her style has been to focus on winning the argument, assuming that the vote will follow.

As Jack Kemp, the leading exponent of supply-side economics, notes, the central question for the entire world in the next century is how to create jobs and wealth. A working, wealthy world will be peaceful. A poverty-stricken world will destroy itself. The fate of the human race in the next century is bound up in the issues of jobs and productivity. The tragedy is that the Left has the wrong answer and the Right has been inarticulate.

Liberals are trapped in an anti-job, anti-economic growth, anti-opportunity cycle because there has been war between the intellectual community and the economic community for a century and a half. During the industrial revolution, rough-and-tumble, aggressive personalities molded the future and grew wealthy by creating jobs and products which made others wealthy also. Aristocratic intellectuals despised these new industrialists and merchants as gauche and lacking in refinement.

Throughout the West, there was a struggle for influence over society between those who made money and those who made books. Among those who made books, both the Right Wing of Ruskin and other conservative intellectuals and the Left Wing of Marx and Weber argued against the irrational, market-oriented, non-intellectual approach of the entrepreneur and the inventor.

This war between the academic and the economic communities led to the growth throughout the West of a Socialist-Communist-American Left critique of business and the free market. Whether it is John Kenneth Galbraith or François Mitterand, there is a strain of left-wing thought which instinctively favors centralized planning and redistribution of wealth.

The tragedy of the left-wing analysis is that it is blind to reality. The simple fact is we know how to create a strong, growing economy which in turn creates jobs and produces wealth.

The greatest economic success stories in the last thirty years are all in the East Asian Littoral running from Japan, through South Korea, to Taiwan, and on to Hong Kong and Singapore. The argument that these successes are a function of Oriental culture is pure and simple racism. The Chinese aren't more frugal or harder-working than any other group. The Japanese aren't racially more inventive than anyone else. The simple fact is that kids who take homework home and do it will eventually outperform kids who ignore education. Adults who work hard and cooperate with each other will eventually outproduce workers who loaf on the job and hide behind a labor-management civil war. A society whose citizens save 19 percent of their income will have much more money to build factories than a society whose citizens only save 6 percent of their income. All these steps are open to Americans.

Simple misperceptions have prevented welfare-state politicians from learning the obvious lessons of East Asian success: 1) American arrogance after the great successes of World War II led a generation to believe it could break the rules and still succeed; we behaved like a sailor on a weekend binge, and the very arrogance which led to that behavior makes it hard to see beyond our cultural pride and learn from others; 2) the welfare-state image of the citizens as victim makes it impossible to learn the lessons of free people and free markets; freedom is incompatible with the victim mentality; it is impossible to design a dynamic, growth-oriented society within a framework of victimization; 3) the welfare state is explicitly committed to policies which cripple economic growth; being soft on crime is as harmful to small businesses as are high interest rates and high taxation; a welfare system which discourages marginal workers from working does more to kill businesses which are labor-intensive than any government regulatory policy; emphasis on the importance of the bureaucrat and the power of the politician drives people out of small businesses by making their work and risk psychologically and socially unrewarding.

Supply-sided economics and politics have suffered grievously since 1981 from too narrow and shallow an approach to the debate with the Left. The Left has demanded that supply-siders talk in econometric numbers when there is no supply-side econometric computer system. Running supply-side ideas through computers based on welfare-state assumptions will *by definition* discredit those ideas.

The best guides to the supply-side approach are Jack Kemp's *An American Renaissance* and Jude Wanniski's *The Way the World Works*. As they explain supply-side reasoning, the culture of success and the psychology of success play a large role in how rapidly we create new jobs. The argument for honest money is psychological rather than economic: people who believe that their government will steal from them through inflation behave differently from people who believe that their government will use an honest, fixed standard to allow them to keep the fruits they harvest.

The welfare state sees men as units to be shaped by abstract numbers. Operating under such an assumption, econometrics is a good policy tool because it forecasts how we will respond in the aggregate to the central bureaucratic state. An opportunity society would argue that man is an organic child of God, that economics and economic numbers are simply a reflection of that behavior.

Because the "supply side-demand side" debate and the larger "opportunity society-welfare state" debate are struggles over the nature of reality, it should be possible to settle the debate by looking at reality. The reality that I see overwhelmingly fits the opportunity-society model and discredits the welfare-state model. People can refuse to look at data which challenges their deepest beliefs, although, carried far enough, we consider that a sign of insanity. Nonetheless, the case for rethinking Liberalism is now so overwhelming that the next generation of intellectuals must almost certainly do so.

Business Management and the Opportunity Society Model

One proof of the negative impact of the welfare-society value system has been the gradual bureaucratization of the American business community. It is a truism that the more large corporations come to resemble governments, the less successful they become. Those corporations which behave most "rationally," in the most centralized and structured manner, which treat their employees as mechanical objects to be managed, and which are the least oriented toward traditional values, are the very corporations most likely to be in real trouble.

Successful corporations see their employees as holistic, organic beings to be nurtured and grown, emphasize core values—honesty, hard work, and discipline—decentralize so that many human-sized units can take risks and grow through responsibility, and focus on technical competence as well as future-oriented, longer-term models of profitability.

The values of successful corporations are a key part of the opportunity-society movement. The fact that a welfare state will eventually dominate the host culture to a degree that even industry will succumb to its incorrect teachings is one more indication that we need a cultural revolution which will develop a new governing majority. The difference between the two models of management is brilliantly explained in Peters' and Waterman's *In Search of Excellence.*

The Individual Entrepreneur and the Opportunity Society

History, in a free society, is largely biography. The Marxist interpretation of history, which focuses on mass behavior and mass change, is simply inaccurate in a nation which emphasizes personal freedom and personal choice. Even within very large structures, it is amazing how much powerful personalities can change entire organizations.

The creation of new technology and the development of new wealth remains largely the function of strong individuals. While large bureaucracies and large corporations can keep the ball rolling, it is in strong individuals—or at most a very small team—that we find the secret of creating the ball and getting it rolling in the first place.

Life, in fact, does not occur "writ large." There are general rules and general principles, but life is still lived at a personal level. It is as individuals and families and neighborhoods, as small businesses or as small units of big businesses, that we identify our lives and our world. In my training as a politician, I learned far more from the biographies of Jefferson, Lincoln, the two Roosevelts, and others than I ever did from abstract studies of political theory. It was the daily working of politics—life in the campaign rather than life with the political scientist—that taught me my trade.

Successful persons seem never to explain their lives

to me in terms of econometrics, discount rates, or the "welfare state-demand side" economists' vision of reality, but in terms of inner values, of the power of traditional beliefs to protect and nurture through hard times, of the need for discipline and persistence, and of the capacity of dreams to change behavior.

People who want to create jobs should study the lives of successful entrepreneurs, such as the autobiography of Eddie Rickenbacker who built Eastern Airlines into a giant.

In my Congressional district in Georgia, as in every area of the country, there are men and women who, as successful entrepreneurs, have made the world around them more successful. These people are our most valuable learning-resource on the subject.

Lessons learned at the national, corporate, and individual levels demonstrate that the welfare state does not work. But there *is* an approach that *does* work.

The opportunity society calls not for a laissez-faire society in which the economic world is a neutral jungle of purely random individual behavior, but for forceful government intervention on behalf of growth and opportunity. We must encourage the production of new wealth, new ideas, and new inventions, as we must bias our society in favor of entrepreneurs and hard work.

Government in an opportunity society will need a new focus and a new set of indicators to measure the number of new entrepreneurs starting business and to design tax laws, rules, and regulations which encourage individual initiative.

All Americans should have the chance to hold jobs and to create a better world for themselves and their families. A half-century of rule by the welfare state has not achieved that great dream. It is time to learn the values and habits of success.

CHAPTER 9: TRADITIONAL VALUES AND THE AMERICAN DREAM OF A GOOD LIFE

Remarkable things have happened to traditional values in my lifetime. When I was a boy, the movie *The Moon Is Blue* was banned in Boston for using the word "virgin." A group in southern California was trying to keep *Tarzan* out of public libraries because Tarzan and Jane had not married; nothing overtly immoral occurred in the book, but the group thought that cohabitation without sanction by clergy was, in and of itself, immoral and, therefore, to be banned. We prayed in school. Hugh Hefner had not yet founded *Playboy* magazine and "bunnies" meant only animals.

When I was young, people smoked corn silk, but cocaine was not a middle-class option. You could walk safely in your neighborhood. Elderly women felt comfortable sitting on their porches in the evening.

Traditional families and traditional lifestyles were so universal that not only was there no debate about alternative lifestyles, there wasn't even any discussion about the concept of alternatives.

I am not recounting these memories simply to recall "the good old days." Rather, I am trying to show how sudden and how radical has been our shift in expectations of the very nature of our personal and social lives and values.

My grandmother would have been surprised by the space shuttle and personal computers, saddened by the wars in Lebanon and Afghanistan, pleased by heart and kidney transplants, and concerned by the rise in illiteracy.

But the change in everyday life would have absolutely staggered my grandmother. She would have hated living in fear and locking herself in at night. She would not have wanted to worry about her grandchildren becoming drug addicts. I doubt that she would have been comfortable in a society in which alternative lifestyles are more acceptable among our elite than church-going and temperance.

Most Americans now think that the pendulum has swung too far. While they don't want *Tarzan* banned from the library, they are not very pleased with sado-masochism or child pornography. While they don't favor control over what adults do in their private bedrooms, they generally don't believe homosexual couples should adopt children or teach in schools or that homosexuality is an acceptable alternative lifestyle. While they want fair trials and professional police forces, they want criminals locked up and they want safe neighborhoods.

By an overwhelming margin, the American people tell our pollsters that they want real change in their government and society with regard to prayer in school, drugs, crime, and pornography. When I hold town-hall meetings or visit neighborhoods in Georgia, I am constantly reminded of the depth of feelings about traditional values among my constituency.

Yet the Washington elite has ruled that protection of our traditional values is not a legitimate government issue. The Washington elite is almost totally focused on economic issues, and it scorns efforts to raise moral and ethical questions. Despite the vast popular support for stronger sentences for criminals, tougher efforts against drug dealers, and voluntary prayer in school, there is no sympathy in Washington for making these matters important in our national agenda.

Shortly after the 1984 State of the Union Address, I was invited to attend a Spirit of America rally in Atlanta and speak to it just before President Reagan did. I had been very impressed with the President's point that Congressmen were allowed to open their daily session with a prayer but their children were prohibited from praying in school.

The Spirit of America rally had been jointly sponsored by the Chamber of Commerce and the Amway Corporation, so it was an economic rather than a religious gathering. I told my hosts I was going to mention the prayer-in-school issue and suggested that they pay attention to the response.

There were fifteen thousand people in the Atlanta Omni center. When I quoted President Reagan's remark calling for voluntary prayer in school, they rose to their feet, applauding and cheering. A few minutes later President Reagan arrived and, in his speech, made the same point. The crowd promptly once again got to its feet and spent a full minute applauding and cheering his remarks. One of the Chamber of Commerce leaders who had scoffed at the importance of prayer in school turned to me and commented, "You may have something. This seems to be a lot more powerful concern than I'd realized."

Despite growing evidence that the American people are hungry for spiritual guidance and traditional values, the national elite rejects the legitimacy of these issues in the legislative and political arena. Millions of Americans watch or listen to religious broadcasts each week; thousands come to Washington to ask for prayer in school; opinion polls demonstrate an overwhelming desire for tougher laws against criminals and drug dealers. These items are still not on the elite's agenda.

The bias of the welfare state against traditional values came clear to me as I prepared to raise the school-prayer issue in the 1984 session of Congress. In one interview, a *New York Times* reporter promptly snapped, "Yes, but isn't that just a smokescreen to avoid serious issues like the budget or the economy?" From the perspective of his value system, it was simply inconceivable that prayer was something serious.

A few weeks later, several of us had a chance to outline our legislative strategy on school prayer for the Washington news staff of a major television network who could not understand why school prayer was on our agenda: their world view could not encompass moral and spiritual matters as serious issues.

It is a comment on the triumph of economic and quantitative analysis that spiritual and religious concerns seem so foreign and so irrelevant to the present elite. It explains a great deal about why they would underestimate the spiritual toughness and fanaticism of Ho Chi Minh and the Communists in Vietnam, why they would think you could buy an agreement with a fundamentalist Islamic reformer like Khomeini, why they would misunderstand the importance of spiritual and emotional commitment in rising from poverty or developing an educated younger generation.

By emphasizing the physical and economic, our intellectuals in effect misreport the equally important characteristics of spiritual, moral, and personal life. Gary Will, in his study of the Declaration of Independence, *Discovering America,* described George Washington as a man whose very being carried a kind of moral power, who dominated a room not only by his great size and strength but also because he was a person whom others venerated. Modern intellectuals discount the power of personality, of character, of pride, and of spiritual and moral leadership to such a degree, Will asserted, that no 20th-century American historian can explain Washington's role in shaping America.

This emphasis on the power of material goods and systems over the power of the mind and spirit has led America repeatedly astray. We have better school buildings, but more children cheat on tests. We have fine shopping centers, but more people get mugged in them. We have more freedom to read pornography, while we seem to find less satisfaction in our personal relationships.

The collapse of our moral and spiritual system is driven home to me every time I talk to a high school or college class. I always ask how many students know someone who cheats on tests. Most hands go up.

Think of it: our young people learn together to

cheat themselves and their society, and they cheat on such a scale that students almost never turn in those contemporaries who break the rules.

The answer is not better test monitors or more vigilance from teachers. A free society cannot survive by increasing its capacity to coerce people.

The crisis of cheating in the schools, the underground economy which cheats our tax system, and the waste and theft in Pentagon procurement all have the same core problem: free society must rely on ethics and honesty for self-enforcement of its rules. If its spiritual and moral values decay enough that only policemen can enforce the rules, then freedom is in danger of disappearing.

A free society must focus on ethics and honor, not on legality and enforcement. Yet the base of ethics and honor must be some commitment to a world which values the spiritual and the moral. In a purely economic world, the tax cheater and the cheating student *are* both winners. If they don't get caught, they win; and—in a decaying welfare state—if they do get caught, they simply plea bargain for mercy. In both cases, a purely economic analysis would encourage behavior destructive to society as a whole but rewarding to the specific individual.

The need for a moral revival is a major factor in my commitment to voluntary prayer in school. It is quite true that there might be some peer pressure to pray even with a voluntary school prayer. When I was asked about that while addressing a high school assembly, I answered, "In an age where there is peer pressure to drink, to use drugs, and to sleep with someone, maybe it wouldn't be a disaster to have some peer pressure to pray for deliverance from alcohol, drugs, and sex."

Horror stories from the Left suggest that America is in danger of becoming a theocracy or that religious persecution is just over the horizon. The American Civil Liberties Union and its allies cite a very restricted and narrow reading of the Constitution, ignoring historical context, to argue that we are protected from religion.

In fact, the central issue involving traditional values is whether the Constitution protects freedom *of* religion or whether the Constitution provides for a secular state which is free *from* all religions.

The ACLU extreme is a major reason there is now a popular backlash on this issue. Students in one of the high schools in my area are forbidden by a Federal Court order from wearing a T-shirt that mentions their church. They can wear T-shirts that use obscenities (protected under the First Amendment freedom of speech). They can wear Playboy Club T-shirts. But a Federal judge has invoked the power of the state to insure that they don't contaminate their fellow students with the name of their church.

In Michigan, a father has his sons taken from him because he has been educating them at home. No one questions that the children can read, write, and do math. The issue is whether he has the right to raise his children according to his values. He and his wife are deeply religious and believe the secular state is corrupting their children with improper values and poor education. In 1619, this couple would have been Pilgrims, fleeing to America for the opportunity to worship as they believed.

In Nebraska, six fundamentalist Christians were jailed for three months in a fight with the state government over how to educate their children: they wanted their children to attend a Faith Christian School which did not have state-accredited teachers. They were willing to have their children tested to prove that they could pass any level of learning proficiency which the state thought necessary; however, Nebraska's state law is accreditation-oriented and the state's response was to jail these men who were willing to guarantee educational attainment for their children, but in a religious atmosphere that fit their values.

For a generation, the American people have allowed a Liberal elite to impose radical values and flaunt deviant beliefs. For a generation, the values which were written into the Declaration of Independence (which asserts that our Creator endows us with certain inalienable rights) have been eroding; we have carried money saying "In God We Trust," while politicians and the ACLU have driven God out of our public life and institutions.

Today we see a country regaining its morale and rebuilding its sense of the importance of spiritual and moral values. The scientific marvels which allow us to see the Earth from outer space are actually increasing interest in spiritual and moral life. Our age of opportunity will involve spiritual and moral, as well as physical and scientific, opportunities.

This may be the hardest forecast for the Liberal elite to swallow. Our intelligentsia and news media will find it easier to believe in lunar bases and supercomputers than to consider the possibility that spiritual and moral issues may be re-emerging as a central theme in American life. Yet I am absolutely certain that most of the generation of Americans which has tried life without God, without prayer, and without moral and spiritual commitment has found such life to be empty and unsatisfying. And before too long, our leaders will have to bow to the will of those they govern.

CHAPTER 10: THE DILEMMAS OF AMERICAN FOREIGN AND MILITARY POLICY

I was born in 1943 during an age of extraordinary American success, when American power was inexorably defeating its enemies in a global war; I entered school the year American power leaped over the Berlin blockade to airlift survival to a beleaguered people; my high school years were sheltered by Eisenhower and an American economic and military power that seemed to tower over the planet.

By the time I went to college and graduate school, things had begun to change. First, the Bay of Pigs marred my freshman year in college with American defeat and embarrassment. My sophomore year was shaken by the fear of nuclear war inherent in the Cuban missile crisis. From Kennedy's assassination in 1963 on through the decay of Vietnam, the assassinations of Martin Luther King, Jr., and Bobby Kennedy, to the humiliation of our rout from Saigon, America failed again and again to understand either the limits of its national will or the uses of its extraordinary power.

America today is increasingly incoherent in its foreign and military policy. We act decisively in Grenada and irresolutely in Lebanon. Congress overmanages minor issues and ignores the need for fundamental reform in our diplomatic and military establishments. We have spent too little on military training and readiness while wasting too much money on expensive active-duty forces where reserve and guard units might well have served.

My lifetime has been a period in which America reached a peak of clarity, power, and decisiveness and then gradually began to decay. Our ability to manage our foreign policy and military bureaucracies has decayed in parallel with our declining margin of strength and security. While our margin for error shrinks, we find our present personnel and systems increasingly likely to make errors.

If we do not find a way to reform our defense and foreign policy systems in the next decade, we run a real danger of suffering a military or foreign policy disaster of the first order.

People tend to think that the world is stable and that alliances are permanent. Nothing could be farther from the truth. Alliances can change overnight.

For the past forty years, Western Europe, Japan, and the oil kingdoms of the Middle East have been allied with us because they have believed that such an alliance was in their interest. The Latin American countries have grudgingly taken a back seat and accepted our neglect at remarkably low cost to our defense and foreign policies. We have encouraged decolonization in Africa and Asia at considerable cost to our European allies and some advantage to ourselves.

Now that period of an American-dominated world is clearly ending. Our economy is not as strong as it was. Our military is not as powerful as it was. Ally after ally has complained about the zigzag, undependable nature of American policy. At some point in the near future we may seem too undependable and dangerous an ally to risk.

Any country which looks at the fate of the South Vietnamese and Cambodian governments, at the fate of the Shah, or at the fate of our recent allies in Lebanon, may start asking whether it really wants to risk being the next international headline.

The degree of European concern over this issue was brought home to me when a senior German general commented on the shifting of an American brigade to northern Germany near the port of Bremerhaven. He noted that the German press had characterized the move as a step toward securing a port for an American army to retreat from Germany in case of a war.

"Like Dunkirk?" I asked him.

"No, like Saigon," he replied with a bitterness and a venom which still shakes me seven years after the encounter.

It frightens our weaker allies to see us indecisive, irresolute, and defeated, when they see us shift from aggressive bluster to passivity without discernible warning.

The American press and many American politicians would agree with much of this analysis, but then focus attention to specific personalities—Carter vs. Reagon, or Brzezinski vs. Kissinger vs. McFarland at the National Security Council—or describe the sudden political whirlwinds that race through Congress leaving policy in wreckage behind them.

Yet this surface analysis of personalities and temporary events is precisely part of the problem. The major weaknesses in the American approach to foreign policy are not questions of who is in charge or even of who is reporting on who is in charge.

The major problems in American foreign policy are problems of intellectual framework, historical memory in the American body politic, institutional structure, and the style and rhythm of our news media and politics—problems which will almost certainly destroy the Western Alliance if we do not address and solve them.

This topic requires a book of its own. For the moment, this chapter can only survey problem areas in which conflicts exist that must be resolved if the Western Alliance is to survive and America is to remain free, prosperous, and safe.

The Crisis in the Intellectual Framework of American Foreign and Defense Policies

The Right and Left Wings of American political and intellectual thought are in conflict about the nature of the world. The Right sees a world of enormous danger in which military strength and constant vigilance are necessary just to survive. The Left sees a world that could be one of human brotherhood and decency if only America would take the first risk and act peacefully.

The Right focuses its attention outward. It is always urging more military strength and intelligence agency activity to offset the dangers of the Soviet Union and its allies. The Left focuses its attention inward on the United States, noting the degree to which our strength becomes a potential threat to others and our firmness is interpreted as bellicosity.

The Right warns of the dangers of tyranny being imposed through military defeat. The Left warns of the danger of mass annihilation if we use our military strength.

The difficulty for American political leaders and for American foreign policy is that both ideological wings have a great deal of truth on their side.

The world is just as dangerous as the Right suggests, but our true goal should be human peace rather than a fortress America. Those conservatives who suggest that the Soviet leadership is determined to dominate or destroy us are correct, but so are those Liberals who warn that an all-out nuclear war would detroy civilization as we know it.

Our generation gave mankind both Auschwitz and Hiroshima. The Germans and the Soviets perfected the modern, technologically advanced totalitarian state with terror, torture, concentration camps, and chemical warfare. We and others created nuclear weapons and the prospect of mass annihilation. All humanity hangs in the balance between totalitarian tyranny and technological annihilation.

Since many conservatives scoff at the dangers of annihilation and many liberals scoff at the dangers of totalitarianism, it is worth examining both threats in their full horror.

Consider first the dangers of nuclear war. With the explosions at Hiroshima and Nagasaki, the genie was truly out of the bottle. Mankind now has the capacity to inflict destruction on a scale which would, at minimum, disrupt civilization as we know it.

It is important to recognize that disasters of comparable scale are not unknown to human experience.

The Mongol destruction of the Islamic world around Baghdad was incredibly complete, leveling that civilization to an extent from which it has not recovered today, some nine hundred years later. The Spanish destruction of Aztec and Inca civilizations was complete. The population loss among American Indians was probably the equivalent of a medium-sized nuclear war. The Black Death killed ⅓ of the population of Europe during the 14th century in a cataclysm that came close to that of a medium-sized nuclear holocaust.

I don't recite these examples of horror so that you will relax and decide that we will survive a nuclear, biological, or chemical cataclysm. Just the opposite. I want to assure you that such a disaster is possible and that it would be horrible.

Remember, in a nuclear war none of us would play the Mongol or Spanish role. There would be massive devastation with a very real likelihood that everyone would be a loser. Even a small and limited nuclear war between the United States and the Soviet Union would kill more Americans than have died altogether in all our wars.

President Eisenhower, after pondering the prospects of a nuclear exchange, said that he would not want to go to a bomb shelter because he would not want to survive and have to deal with the pain and the horror of the aftermath. Our nature as a species, however, seems to be that most of those who survived would want to continue surviving.

The problems of mass destruction are about to be compounded by the proliferation of the technology of mass death. Whatever difficulties and fears we have had in the past with a bipolar confrontation involving nuclear weapons will be dramatically heightened in the next generation.

The technology for killing a lot of people has been spreading ever since 1945. First, the United States, then the Soviet Union, Great Britain, France, and China publicly acquired weapons capable of destroying cities. More recently, India has exploded an atomic device. We have reason to believe Israel, South Africa, and Sweden either have or will soon have nuclear weapons and that Pakistan is working on a nuclear weapon to gain parity with India. The Israelis were worried enough about a potential Iraqi nuclear weapon that they bombed the new Baghdad nuclear power plant to retard that possibility for a few years.

Since the great powers already have over fifty thousand weapons and the technology for creating relatively inexpensive atomic bombs is spreading, the mathematical probability of someone, somewhere, getting a bomb and being willing to use it steadily increases.

Terrorist behavior and attitudes can be examined in situations like the Munich massacre of Israeli Olympic athletes and in the willingness to kill women and children in hundreds of incidents in the Middle East and Northern Ireland. The picture is sobering indeed. Imagine the more extreme elements in any terrorist movement with weapons of mass destruction. It is a prospect likely to gray the hair of any reasonable person.

Yet our problem will not come only from terrorist, illegal organizations. There are bandit nations willing to operate outside the tradition of modern international behavior. The three most obvious current bandit governments are North Korea, Libya, and Iran. The leaders of all three countries are inner-directed and likely to do what they personally decide is appropriate. All three leaders have proven themselves risk-takers willing to subsidize terrorist organizations and willing to kill innocent people in the pursuit of their goals. The thought of them having nuclear weapons is daunting indeed.

The American Sixth Fleet would have a whole new set of problems staying in the Mediterranean if Quaddaffi had even modest nuclear weapons available. Furthermore, we must remember that it is only in the West that we focus military power on military engagements. There is every reason to believe that Middle Eastern ideologues will strike at the American heartland rather than at our military power if we threaten them directly.

While there are some dangers to the United States from the proliferation of weapons of mass destruction, there is much greater danger to the human race in general. It is quite conceivable that Pakistan and India would use nuclear weapons on each other, that the PLO will find some new biological or chemical weapons to use on Israel, or that Iran will buy or develop a terror weapon to re-establish Persian hegemony over the Persian Gulf in the name of a fanatic Islamic revival. These will be the realities of our children's and grandchildren's world.

We live constantly in the shadow of the holocaust which could destroy civilization. I have been convinced that we live on the edge of disaster ever since I was fifteen years old. When my dad was stationed in France for the United States Army, we visited a friend of his at Verdun who had been drafted in 1941 from a small town in Pennsylvania and sent to serve in the Philippines. One year later, he was part of the Bataan death march and served three years in a Japanese prisoner-of-war camp.

We spent evenings talking about the plight of a young man who was uprooted from his family, who suddenly found himself in a strange world, under the domination of guards from a foreign culture, being forced to eat inadequate quantities of a formerly unthinkable diet under living conditions which were inconceivable to him only months earlier.

During the day we toured the battlefields of the greatest killing ground of World War I. There, in 1915, the French and German armies collided in a struggle in which each was determined to bleed to death rather than yield. The Germans were convinced that victory would break the morale of France and end the war; the French agreed with the German analysis and were determined to hold their fortress at any cost.

Those decisions created a conflict in which hundreds of thousands of men died in a small valley overlooked by mountains. Even today, few sights have quite the impact of walking through fortifications dug into mountains from the rear so they can house the artillery which spewed death from the front. The French poured enough concrete into these forts to have rebuilt much of France. Yet the striking thing, in 1958, was that many parts of the town of Verdun were still shell-scarred and ruined from that battle forty-three years earlier.

It was hard to imagine that war could be so ferocious and so all-consuming, that the very terrain would not have recovered in that length of time. Any doubts I might have had were dispelled when we visited the Ossuary.

The Ossuary is a large building overlooking the battlefield of Verdun. Its glassed-in basement holds the bones of more than 100,000 bodies which could not be identified after the battle because they had been blown apart by artillery and mines and left rotting in the fields during the nine-month battle.

Today, twenty-five years after I visited Verdun, I can still feel the sense of horror and reality which overcame me then. I have never been back, yet Verdun is as real to me as if I had visited it last weekend. It is the driving force which pushed me into history and politics and molded my life.

I left that battlefield convinced that men do horrible things to each other, that great nations can spend their lifeblood and their treasure on efforts to coerce and subjugate their fellow man. I was absolutely certain that what had happened before could happen again.

Yet, much as I might recoil in horror from the evil and death of Verdun, I could not bring myself to favor disarming unilaterally. For the other part of our 20th-century experience has been the horror of the death camps and the Gulags.

Winston Churchill once said, "War is horrible, slavery is worse." The more I study Auschwitz, Dachau, and their equivalents, the Gulags of the Soviet system, the more I become convinced that there are some things even more horrible than Verdun.

American intellectuals, Liberal politicians, and the American news media find it difficult to accept that the Soviet system really is totalitarian, ruthless, inhumane, and relatively unchanging.

George Kennan's "containment" thesis posited that, if we blocked Soviet expansion, their system would gradually mellow and liberalize. This view is, in retrospect, one of remarkable naivete. The Soviet system is the linear and even more horribly efficient descendant of the Czarist system before it. The KGB's security police is the technologically more powerful direct descendant of the Czariat security police. The Soviet willingness to imprison artists and intellectuals like Sakharov and Solzhenitsyn is the precise imitation of the Czarist willingness to imprison Dostoevsky and other writers.

The extraordinary staying power of national styles and national cultures is remarkably illustrated by two books written by French travelers in the 1830s. The more familiar book for Americans is DeTocqueville's *Democracy in America,* but De Custine's *Journey for Our Times* is even more fascinating.

De Custine was an embittered French aristocrat whose father and uncle had both been guillotined during the French Revolution. He hated constitutional governments and sought to defend the Russian aristocracy and autocracy. But De Custine returned from Russia a passionate believer in Western constitutional governments and firmly opposed to bureaucratic oppression.

General, then Ambassador, Walter Bedall Smith (former chief of staff for Eisenhower) found De Custine's book in a Moscow used book store. He had it translated into English because he considered it the most accurate portrait of Stalin's Russia yet written. He explained in the forward to the American edition that things had changed so little in Moscow that changing the names and dates would have allowed him to send large portions of De Custine's work to Washington as reports on Stalin's government in the late 1940s.

American politicians are fond of noting how accurate DeToqueville's observations about our country were and how valid many of them still are. His eastern-travelling counterpart was apparently equally accurate, and his observations remain equally valid.

We must then confront the reality that the Soviet Union is not likely to become more democratic and more open any more than we are likely to become more totalitarian. The Soviets are no more likely to accept free debate than we are to accept security police.

Americans tend to forget that there is a very powerful Soviet governing class which profits from the current system. These are the people who have special passports, who can shop in special stores, who get their children into special universities, and who can take special vacations.

The Soviet governing class knows full well how miserable life is for the Soviet worker. Among ordinary Soviet citizens, alcoholism is widespread, life expectancy is low, and consumer goods are very scarce. The rulers are unlikely to share their bounty among the Soviet underclasses. The security police and the party stand shoulder-to-shoulder in watching for any softheartedness which might creep into the existing system.

We must expect the Soviet system to survive in its present brutish form for a very long time. There will be Soviet labor camps and Soviet torture chambers well into our great grandchildren's lives: great centers of political and economic power have enormous staying power; Czarist Russia lasted through 3½ years of the most agonizing kind of war; the Nazi state did not collapse even when battlefield defeats reduced its control to only a tiny sliver of Germany.

We must therefore assume the Soviet Union will survive as a dangerous totalitarian state. Since a deliberate major war is unthinkable, there is no practical American policy for changing the Soviet regime. Our pious hopes of the last half-century have culminated in the shooting down of a Korean airliner and the Party Chairmanship of yet another member of the Soviet hardline gerontocracy.

If we cannot fundamentally change the current

Soviet regime, then what can we do? Our goal must be to survive in freedom and prosperity without war.

A century from now, the free Western Alliance will have grown economically and technologically beyond the Soviet Union's capacity to compete. Conceivably, if we retain our morale and our commitment, the Soviet elite may eventually tire of repressing its own flesh and blood; but that is a hope, not a basis for policy.

However, for us to survive the next century, we must resolve the problems outlined in this chapter. The first is the intellectual split in American foreign policy between the realism of the Right and the idealism of the Left.

1. *The Conflict Between Realism and Idealism*

We cannot continue to lead an alliance if our leaders zigzag back and forth between the politics of force and strength on the one hand and a Wilsonian idealism of unilateral restraint and withdrawal on the other.

The Right must address the issues of creating a less conflict-ridden, less dangerous world.

The Left must confront the realities of chemical warfare in Laos, the Soviet invasion and brutal repression in Afghanistan, Cuban troops in Africa, and Soviet-inspired tyranny throughout the Warsaw Pact nations.

Freedom can be preserved and massive violence avoided only by following the basic rules of history. I believe passionately in Madison's injunction that "knowledge will forever govern ignorance," and I am afraid that my Liberal friends are remarkably ignorant about history and about the track record of human passion and human violence.

I want my children and grandchildren to live safely and in freedom. I am convinced it is possible for us to provide a dynamic expanding society with hope and opportunity which will lead to a more positive relationship between human beings. However, I also believe that managing the transition from the world we are living in to the dream we could live in will be very difficult.

The symbolic, simplistic efforts of the Left are legitimate as emotional expressions of their fear of war. It is reasonable for anyone who has thought seriously about the horrors of nuclear, chemical, or biological conflict to say, "Stop; no more; do anything to avoid it."

However, simply waving the totem of a unilateral nuclear freeze or talking piously about our love of peace may accomplish just the opposite of our intentions. Neville Chamberlain's passion for peace increased the likelihood of World War II. The passive American reaction to the Japanese invasion of China increased the prospects of future war. Agressive and expansionist movements and nations are encouraged when their opponents are reasonable, understanding, and willing only to talk. They will then exploit every opportunity to expand their power and their capacity for evil at the expense of decent people.

I am convinced that there is an alternative to both the naive utopian vision of the Left and the bland lack of interest in many of these issues on the Right. Too many conservatives shrug off the danger of nuclear war. Too many conservatives fail to think deeply enough about the need for a new diplomacy and a new world framework in which to develop a peaceful society for all human beings.

If the Left has been too glib and too eager in its search for immediate solutions to a difficult problem, then the Right has been too unimaginative in its failure to see that the world really is changing and the problems of violence are really of a different magnitude than they were before Hiroshima and Nagasaki.

The split between idealism and realism runs deep in the American political, intellectual, and news-media elites. Our foreign policy and defense bureaucracies are buffeted back and forth as first one imperative and then the other dominates our policy-making.

From dealing with terrorism through low intensity conflicts to the danger of large-scale conventional war, the American system is confused and indecisive in its orientation. One impulse drives us toward involvement, military strength, and some risk-taking, with the consequent possibility of having Americans killed in action or by terrorists. The other impulse pursues negotiations and conciliation at the risk of humiliation and avoids violence at the risk of breaking our treaty and moral responsibilities.

In many ways, the intellectual and moral dichotomy in the American foreign and defense policy area is comparable to the confusion an adolescent feels when torn between varying desires and varying urges. It is vital that, like an adolescent, we as a nation develop a unifying systhesis out of our current thesis and antithesis.

Until we can shape a long-term intellectual framework for foreign and defense policy, we will continue to solve tactical problems with an inadequate strategic and visionary policy and framework. As long as we keep making tactical zigzags, we will continue to decay as an alliance leader. Therefore, the first imperative of a successful revitalization of American leadership for the free world is a new vision that combines the realistic demands of survival in a dangerous and often unreasonable world with the idealistic impulse which at its best is the very core of the American dream.

However, creating an intellectual basis for planning and implementing our foreign and defense policies is not in itself enough to secure our survival. We must still come to grips with the fact that our historical memory is so different from the contemporary behavior worldwide that relying on our national understanding of human behavior may get us into very great difficulties.

2. *American History as a Misleading Guide to Survival in a Dangerous and Often Unreasonable World*

All our intellectual and theoretical efforts will be in vain if the basic historical memory of the American people remains unchallenged. Our past experiences were so much more peaceful and our ability to survive without military preparedness were so great that it literally threatens to kill our chance for survival.

In many ways the Soviet vision of a dangerous world in which enemies will attack you and destroy your cities is a much more realistic and accurate assessment of the traditional patterns of history: the Russian historical experience in a world of hostility and violence has been much closer to the norm than our experience of protection by wide oceans and a relatively unoccupied continent with no neighbors powerful enough to threaten us. They are clearly not like us and do not share our memories, our assumptions, or our values.

As the world has crowded in on our borders and the potential for violence against our people has increased, we have had a hard time coping with the changing realities. We have paid considerable costs as a people to avoid changing our positive and relatively naive historical world view. This resistance to uncomfortable information and unpleasantly changing realities is normal, but dangerous.

People do not like to be told bad news. We do not like to be reminded that there are objective realities which require us to subordinate our egos and forego our pleasures. I love ice cream: as often as I can, I lie to myself about my weight.

In other words, we are all capable of cognitive dissonance, that is, the ability to selectively deny information which threatens our basic beliefs. A person who believes in hard work may reject stories about people getting lucky: relying on luck would threaten his or her value system. Conversely, a person who believes that the world is random and individual effort doesn't count for much would reject stories about people who have struggled and sweated and worked their way to success.

There are several patterns in American history which have shaped our Liberal elite's view of reality and which now threaten our very survival.

First, we have been blessed as a nation by being a long way from any major enemy. We simply have no national historic memory, before 1945, of being in a threatening world. Second, our personal histories as communities and groups have been relatively peaceful. The only civil war we have fought was between two large regions. We have no tradition of the kind of ruthless neighbor-against-neighbor brutality which Northern Ireland, Lebanon, and Cambodia have experienced. Third, our national ethic is devoted to the notion that reasonable people can negotiate peacefully and find a way to accomodate each other. Our entire cultural-social-political-economic system is based on the idea that there are common meeting grounds which allow people to live together despite ethnic, religious, class, or racial differences.

With the tragic exceptions of segregation for Black Americans and of long-lasting discrimination against Chinese and Japanese Americans, we have been a remarkably open society compared to virtually any other in modern history. There once were signs in Boston which said "Irish need not apply," but those signs were temporary. They did not stop Boston Irishmen from becoming mayors or Speaker of the House of Representatives or President of the United States.

Our national historical memory, then, has not prepared us as a people for the normal realities of the planet. Our bias as a nation has committed us to inventing the future rather than studying the past. While that is good, it has also blocked us from learning how constantly and enthusiastically human beings have maimed, tortured, and killed each other. There is a naivete in Americans which blinds us to opportunities to learn. As Santayana warned, we seem a nation doomed to repeat mistakes in dealing with others precisely because we do not value the effort of learning more about them.

We have been surprised again and again by other nations because we refuse to study their habits, their culture, and their history. Five hundred years before Christ, Sun T'zu stated, "Know the enemy and you have won half the battle. Know yourself and the battle is yours." We have a passion for knowing about technology, hardware, and management, but we disdain knowing much about either the capacity of others or ourselves to endure (e.g., Vietnam) or our opponents' techniques and approaches.

Only this willful ignorance can explain our underestimation of the Japanese before Pearl Harbor. Bernard Fall warned us again and again in the early 1960s who Ho Chi Minh was and how long he would fight, but we continued to underestimate the North Vietnamese until they defeated us just as Fall had predicted. We underestimated the Lebanese-Syrian-Iranian-Soviet terrorist connections which had already used vehicle bombs and produced numerous young fanatic volunteers willing to die for their cause, and 241 U.S. Marines died as a result.

Because we reject history as a serious preparation for understanding and operating in the world at large, we find ourselves consistently underestimating how difficult, how intractable, how brutal and violent that world can be. History is powerful precisely because it carries us outside our peaceful neighborhoods and our calm communities. At its best, history can open our minds to possibilities which we would never encounter in our own family or surroundings. The world that *has been* can be again.

Americans in general tend to underestimate the savagery of the world, but Liberals in particular carry the tendency to extremes. Liberals seem to have an ideological block against accepting the notion that there really are dangerous people out there who will do evil things unless they are stopped.

Contrary to the American tradition, there are some people who are actually evil. There was no reasoning with either Adolph Hitler or Joseph Stalin; both were men who respected only strength and had nothing but contempt for weakness.

There really *are* dictators, and occasionally even entire governments, which are the moral equivalent of criminal gangs terrorizing an innocent town. There really *are* sadists and thugs who, given the opportunity, will torture and brutalize because they enjoy doing it.

Since the secular-Liberal world view cannot tolerate the concept of evil, there is no framework for thinking about these objective realities. If you instruct pigs that wolves are an extinct species, or that wolves are pigs' best friends, those pigs will be eaten.

Domestically, the Liberal mindset tends to find excuses and explanations for murderers, rapists, and potential assassins. *Newsweek* prints letters from John Hinckley after he threatened the entire society by trying to kill the President. A San Franciscan who murdered the city's mayor and then pleaded that he was upset by financial pressures, political setbacks, and the chemical impact of eating junk food, was convicted of the lowest possible charge and was eligible for parole barely five years after having deprived a city of its elected leader. After two men in Detroit beat a Chinese-American to death after a barroom argument about automobile imports, the judge sentenced them to a $3,780 fine and three years probation.

The domestic inability of Liberalism to confront and contain evil makes all of us walk at night a little more fearfully. It leads all of us to worry about the dangers of criminal behavior more than we should have to in a free society.

However, the domestic irrationality of Liberalism in rejecting evil has nasty consequences and may increase the number of murders and rapes. It is unlikely, by itself, to lead to our collapse as a country or our extinction as a free society. That is, however, exactly what the same Liberal attitude threatens to do to us in international affairs.

The fact is that we stand on the brink of a world of violence almost beyond our imagination.

From the individual mugger-rapist-murderer-terrorist to the local war (and there were forty-four being fought somewhere on the planet on the day ABC showed *The Day After*), through the terrorists with nuclear bombs, to the horrors of a big spasm nuclear war, we are faced with a range of potential violence almost beyond our imagining.

Just as the comfortable Russian landowner before Stalin could not imagine the horrors of collectivization and the comfortable bourgeois German Jew really could not believe Hitler was serious in his speeches, so it is hard for us to believe that these kinds of nightmares are possible. We keep rejecting information about the world around us because it is too far outside our personal experiences, our historical experience, and our shared general view of the world.

It is this refusal to think seriously about the violence we see each night on television and to develop a new explanation for the world we live in which keeps us at a level of shock and surprise. Watch your own reactions the next three or four times you see really violent news reports about a terrorist or a war or the latest atrocity somewhere.

We are going to have to develop an intellectual split-vision which allows us to accept both the reality of our peaceful neighborhood and the reality of a horribly dangerous outside world. If we don't develop a new sophistication to analyze and deal with the dangers from abroad, we will find those dangers creeping closer and closer to our neighborhood. If we don't learn to take serious precautions and to be honest with ourselves about all levels of violence—from individual terrorist-criminal all the way up to a Soviet-American nuclear war—then we increase the danger that these events will occur.

We already have an example of confronting a terrorist danger and handling it pretty effectively: when people first began hijacking airplanes, the civilized world was stunned; the act of air piracy was something which just wasn't civilized; it was hard at first to imagine how we could stop serious hijackers in our open society.

As the hijackers continued, the public demanded that their ability to fly without fear be safeguarded. Professional security agents began examining airplanes at civilian airports. Within a matter of months, a new industry had been born, a new technology had been developed, and new habits were acquired by the traveling public. Today all of us are used to standing in line and having our bodies examined while our luggage is X-rayed. We accept, as a necessary cost of safety, a level of inconvenience and irritation which we would have found intolerable prior to the advent of hijackings.

Compared to the world we once knew, the current airport security precautions are expensive, time-consuming, and a pain in the neck. Compared to the terrorist alternative, to wondering if our airplane will be hijacked, there are few people who would vote to dismantle the security system we have built.

We will need to develop new systems and new approaches to deal with a wide variety of threats over the next few years. Just as our parents could not imagine the level of security used today at airports, so we can hardly envision the kind of steps we will have to take to thwart terrorists in the future.

In their book *Brittle Power,* Amory B. and Hunter L. Lovins have analyzed the vulnerability of a free society to the dangers of terrorism and come to the conclusion that virtually all our economic planning was done under the assumption that the world would be safe. As a consequence, most American transportation, communication, and manufacturing systems have a variety of choke points which are vulnerable to terrorist activity.

Any terrorist group which systematically exploited the vulnerabilities of our economy would discover that there was remarkably little planning for backup systems in case the normal routes and structures were destroyed. In effect, we have made ourselves vulnerable to any serious group of terrorists who want to paralyze large portions of our economy and society.

On a more political note, we have not dealt with the vulnerabilities of our cities in general and our national capital in particular. In the future, any American effort to exert military force by the use of air and sea power may be met by a determined opponent who decides to use the very openness of a free society as a weapon against us.

We are such a multinational culture that virtually *every* ideological group and ethnic faction in the world has sympathizers somewhere in the United States. A foreign government or movement which decided to attack us in our homeland would probably find at least some allies here who, for money or from sympathy or fear, were willing to collaborate with the terrorists.

Serious students of the Middle East doubt that we could ever expect a Syrian or Iranian government to tolerate the American bombing of their territory without reprisals on American soil. We cannot again expect a one-sided conflict such as we had in Korea and Vietnam, when we bombed and bombarded our enemies but they allowed our country to live in peace.

It would be remarkably easy, in a free society like ours, to find a truck and some explosives and ram the Pentagon, the Washington Monument, or the Capitol. After all, we are willing to use jet fighter-bombers costing up to $25,000,000 to bomb our enemies. Why should we assume that those enemies would not be willing to spend an equal amount in a more cost-effective manner? I will leave it up to your imagination how much damage could be done to any home town for the cost of one F-14 or F-15 aircraft.

Terrorism must be confronted because it is far more likely to have an impact on our lives than is nuclear war. Terrorism is more likely to kill Americans and to challenge our policies than is any other kind of force.

Liberals find it peculiarly difficult to deal with terrorism because terrorism is cloaked in the self-asserted morality of a group that exists outside the law. Liberalism has a passion for fairness which causes it to hesitate. Because Liberalism is still guilty over Western colonialism, it is generally willing to give terrorists the benefit of the doubt and to think of them as underdogs. Finally, Liberals tend to try to find some nice framework of International Law within which to fight a terrorist struggle, rejecting the pragmatic concept of control and suppression in favor of one of negotiation and recognition.

The problem with the Liberal approach to controlling terrorism is that it is almost certain to fail because it involves a fundamental misunderstanding of the nature of terrorist movements and a fundamental misreading of how terrorist movements have been successfully dealt with in the past.

If we do not become practical and candid about the nature of the dilemma we face, we will lose many more men, women, and children to bombings, and we will begin to experience an erosion of civilization here at home. We must develop a doctrine which states clearly American policy toward violence aimed at the destruction of our society. We must take the steps necessary to prove that no terrorist organization can kill Americans with impunity.

The long-term struggle against terrorism will be a dark and bloody one, involving years of vigilant counterterrorism—a level of surveillance and spying that Liberals will call intolerable—and a willingness to strike back with substantial force at the originators of the action rather than the foot soldiers of the terrorist movement.

A free, open society cannot survive by trading violence for violence. If we kill an Iranian extremist every time Iranians kill an American soldier, we will lose the struggle. In the end, no free society can keep pace in enduring pain with a fanatical terrorist organization. We must develop a doctrine which so severely and directly threatens the leaders of terrorist movements that they refrain from attacking the United States because they fear personal consequences. Any other policy is an invitation to a blood bath in which we will certainly be the losers.

The need to develop doctrines and tactics of aggressive counterterrorism goes against the grain of the American historical memory as taught in modern schools. By blotting out the wars against the Indians, the Barbary Pirates, the pacification of the West, and the campaigns against guerrillas in the Phillipines and Central America, it has been possible for the Wilsonian intellectual tradition to dominate—a tradition that argues for a sharp and vivid distinction between war and peace. Liberals dominated by this tradition declare war on a country or are impotent to challenge it; they have no capacity for a long and difficult struggle in the twilight zone of low-intensity conflict.

In the modern world, there are many problems and situations which do not lend themselves to massive, all-out efforts. Nonetheless, if fighting begins, the American people demand large rewards and strong efforts.

General George Marshall said that his biggest mistake during World War II had been underestimating the impatience of the American people. Despite our clear lack of preparedness, Roosevelt was under enormous pressure to fight somewhere before the November, 1942, elections. The public would not stand for a full year of buildup, training, and preparation after Pearl Harbor.

The greatest failure in Vietnam was our own government's underestimation of the speed with which the American people would tire of conflict. That collapse and failure continues to haunt us. Today we cannot sustain fourteen hundred Marines for more than six months in Lebanon without a severe crisis erupting.

The American historical memory of a non-threatening world combines with the moral demand for total war or total peace, creating the illusion that we must be either abstinent or outrageously overwhelming in our application of force.

It *is* possible to use force within reasonable limits to achieve limited goals in limited areas of tension and conflict. Our own history is replete with examples of limited use of strength to achieve our goals. Jefferson daringly used limited forces against the Barbary pirates when we were a relatively weak nation, and used it in the Mediterranean with more success than we have had in Lebanon today. The lengthy cavalry campaigns against the Indians include some which are classic studies in the suppression of mobile guerrilla forces by limited, but *professional*, troops. Theodore Roosevelt used the Great

White Fleet to extend our influence through showing the flag.

We must examine the problems and techniques available to a free society when dealing with foreign defense policy problems if our society is to survive.

However, improving our intellectual framework and modifying our historical memory are still not enough in themselves. Our present foreign and defense policy institutions are too clumsy, too disjointed, too irresponsible, and too unsophisticated to provide adequate policy options to the American government and the American national.

3. Institutional Reforms for Defense and Foreign Policies

American news media have a passion for personality-focused reporting. Virtually every American news source is people-oriented; if a policy succeeds, someone is a hero; if a policy fails, someone is a scapegoat.

People are easier to explain and easier to dramatize than are systems and institutions. News media focused on reaching an audience easily must, of necessity, focus on human struggles and human stories.

This media coverage is reinforced by a recent intellectual-literary fad which translates all policy struggles into gossip and personality-driven efforts for dominance.

There is some truth to this version of a world in which powerful men and women maneuver and manipulate. However, for every major decision based on personality, there are a dozen decisions driven by the structural imperatives of the bureaucracy or profession involved.

The system of the Defense Department matters far more than the personality of any one Secretary of Defense except during the most extraordinary crisis. The structure of the Congress impacts on our national ability to deal with continuing crises far more than any one Congressional personality.

Personality-focused analysis is extraordinarily misleading. It blocks us from diagnosing the real structural illnesses which are crippling this country in its relations with the world.

John Collins of the Library of Congress studied the backgrounds of recent major military and governmental leaders to see how well prepared they were for strategic thinking and planning. Of all our post-World War II Presidents, he found only Eisenhower intellectually prepared to think about long-term conflict. Surprisingly, he found our senior military leaders were not particularly well prepared: most senior military leaders have spent their time either leading troops in tactical formations or serving in bureaucratic positions rather than studying and working at the art of grand strategy. The result is that the very professionals to whom a President is likely to turn are themselves unprofessional on strategic matters.

Eisenhower designed the National Security Council to bring together the State and Defense Departments, to collect systematic analyses and advice for the President, and to develop long-range analysis and planning. The National Security Council is in theory the one body in Washington which links military and diplomatic policies.

However, the National Security Council has become merely the instrument by which the White House attempts to impose its will on the Defense and State Departments during crises. When concerned with almost every hot spot, the State Department will offer one line of policy while the Pentagon will have a different approach to the same area and problem. These two large bureaucracies support their individual positions by selected leaks to the news media and by mobilizing their friends in Congress. Presidents grow frustrated at the bickering and decide that the National Security Council can somehow bring together the bureaucracies and either force them to develop a common policy or impose the President's wishes even if both bureaucracies balk.

Any place on Earth which involves either American interests or American allies is potentially a hot spot. There are likely to be several crisis areas flaring up at any one time.

The news media arouse the public, and the public arouses the Congress. Individual Congressmen then start asking difficult and/or time-consuming questions. The National Security Council then must fight political battles with the Pentagon, the State Department, the legislative branch, and the news media. Since the National Security Council staff is small, it tends to be overwhelmed by crises. As a result, incoherent policies emerge too slowly and too late to anticipate key developments around the world. In addition, the National Security Council has only limited success in managing the bureaucracies of State and Defense. Thus, Presidential policies end up having too little planning, too little coordination, and too many tough public relations and political problems.

The present disarray in American foreign and defense policy is not new or primarily related to Reagan's personality and the competence of his advisors; the system is simply designed for bureaucratic infighting rather than for the serious development of strategy and policy for a world power.

Today the Western Alliance is led clumsily and uncertainly by an America that itself is weakened and frustrated by bureaucracies more interested in achieving their own goals than in implementing the national will. The quality of professionalism and the level of training for international leadership has declined from that of the generation which fought World War II and created the Western Alliance.

We have replaced serious professionalism with a passion for bureaucracy, clear decision memos with turgid, muddled documents. We have permitted the bureaucracies of State and Defense to carve up the zones of conflict-management.

The Defense Department learned from its experience in Vietnam not to get involved without overt and explicit public support, while the State Department is now generally more willing to use force or threaten to use force to secure what the President defines as "American objectives." Since these are rival bureaucracies, they tend to offer conflicting American policies toward large parts of the world.

The lack of clarity in the executive branch compounds and magnifies the demagoguery and sensation-mongering which are always a risk in the legislative branch. Throughout our history, there has been a tradition of uninformed politicians acting in accord with their own short-term interests, even of risking the national interest, in order to improve their own career opportunities. This tradition of Congressional grandstanding and interference includes the Continental Congress, the War Hawks of 1812, Congressional efforts to interfere in the Civil War, and up through, in our own time, the Church Commission which gelded our entire intelligence community so thoroughly in a CIA witch-hunt that our intelligence-gathering apparatus has not yet recovered. Congressmen can cause such problems because policy-making is divided between the legislative and executive branches, and because the executive branch exhibits an appalling lack of professionalism in dealing with the Congress.

It is possible to serve in Congress for thirty years, visit scores of military bases, sit through hundreds of hours of briefings, and retire having learned little of substance about national strategy or the use of military force in a given setting. The Defense Department is now so anxious to placate and appease that it tolerates arrogance, attempts to entertain, and tries to mollify Congressmen. As a result, Congressmen lack serious professional teachers to instruct them in the ways that nations survive.

Sun T'zu opens his classic, *The Art of War,* by asserting that "War is a matter of vital importance to the State; the province of life or death; the road to survival or ruin. It is mandatory that it be thoroughly studied."

Our political, military, and diplomatic leaders do not really study the nature and history of conflict, nor do they spend their time reading up on the history of war, or the history of regional conflicts, or the nature of human interactions when two systems collide on non-negotiable issues.

In all the talks, papers, and reports on Lebanon throughout 1983 and early 1984, the one source I found consistently accurate was Robert G. Neumann, a former American ambassador to several Middle Eastern countries. Neumann clung to three historically based realities: First, as a young man, he had fought in the Austrian civil war in 1934, and so understood that real civil wars (unlike our war between two regions) pit neighbor against neighbor, split families, and leave a legacy of bitterness and bloodshed outside the modern American experience. Second, Neumann appreciated Frederick the Great's adage that "Diplomacy without force is like music without an orchestra," and, realizing that any American presence in the Middle East would be tiny compared to the very large Syrian army and air force, concluded that our diplomatic music would have too small an orchestra to impress our opponents. Third, Neumann knew that the recent history of violence in Lebanon, Israel, and Syria made any American threats endurable and even laughable: short of threatening to annihilate Damascus, there was nothing Americans could do to the Assad dictatorship that the Israelis had not already done; indeed, it was inconceivable that American troops could kill as many Syrians as Assad himself had killed in suppressing the revolt in Hama.

Neumann's analysis was clear and accurate. Given the historical environment and the relative forces, it was obvious that, from September 1983 on, we were running a bluff; though our intentions were honorable, it was pointless to bluff religious fanatics or a ruthless dictator.

All the outpouring articles, speeches, and analyses were less enlightening than Neumann's simple outline. All the State Department figures who rushed up to Capitol Hill with the latest hourly reports, the newest developments, the most unusual tidbits, were misleading, inaccurate, and hopelessly wrong.

I cite this one example because it is typical of the

frustrations of American foreign and defense policy in my lifetime. We have a passion for immediate statistics which mislead. We become bored with the long-term realities that could guide and direct a successful policy.

The history of American willful ignorance—from Johnson on Vietnam, through Carter on Russian intentions in Afghanistan, to Reagan's underestimation of Syria's toughness and staying power—is a long history of current gossip and bureaucratic self-deception. We have ignored objective realities, preferring instead to set up policies which collapse because they cannot be sustained.

We can develop sound foreign and defense policies only if we start with a requirement that our professionals master their professions. When our diplomats are experts in their areas instead of bureaucratic messengers, we will begin to move toward more solid policies. When our generals are masters of the art of war instead of bureaucrats in procurement and paperwork, our military proficiency will improve profoundly.

Only when our professionals master their professions can we begin to design structures that will work. Then we must decide what sort of executive-branch planning and implementation systems are desirable.

At a minimum, we will need closer relationships between the intelligence agencies, the diplomatic agencies, the economic agencies, the military agencies, the news media, and the political structure. There has to be a synergism in which our assessment of what is happening relates to our policies as they are developed and implemented. Both analyses and implementation must be related to the news media and political system because all basic policies must have public support if they are to succeed.

Finally, once the professionals have mastered their professions and have begun to work in systems that are effective and coordinated, those professionals must teach both the news media and the elected politicians. No free society can for long accept the level of ignorance about war, history, and the nature of power which has become the norm for our news media and our elected politicians. An ignorant society is on its way to becoming an extinct society. It is to be the two great centers of political behavior, the news media and the politicians, that we must now turn.

4. Educating Our News Media and Our Politicians

If we are to survive as a free, prosperous, and safe nation, we must educate both our news media and our elected politicians. There is no substitute or short cut for this difficult task.

No free society can expect its professional bureaucracies to develop and sustain policies better than its news media can report or its politicians understand. Indeed, only in a dictatorship can a bureaucracy be strong enough to do what it thinks is right despite its news media and its elected officials.

We are forced, then, to choose between allowing our bureaucracies to decay to the level of the news media which cover them and the politicians who must pay for and approve their policies *or* striving to develop professional competence among our news media and our politicians.

For decades, our bureaucracies have been allowed to decay toward the level of ignorance and amateurism which has always been the hallmark of American politics and is becoming all too common among the American news media. Increasingly, our generals and our diplomats have come to Capitol Hill and to press conferences prepared to answer the questions that would be asked, rather than to give the message that must be heard.

The problem today is that there are tragically few arenas available for professionals to educate either the news media or elected politicians. The adversarial relationship which has become the motif of the current generation of reporters is a barrier to understanding that makes it extremely difficult for reporters to grasp complex problems. The structural bias toward 45-second radio stories and 90-second television stories compounds this difficulty: reporters in the electronic media don't need the full story because they are only going to fill a brief period of time. The result is an increase in the number of stories reported without enlightening context.

There are some encouraging signs of a return to more serious type of broadcast journalism, such as Ted Koppel's *Nightline*, the *McNeil-Lehrer Report's* expansion to an hour, and the efforts on the evening news to move toward 3½- to 4-minute feature stories.

However, one of the problems of reporting on the modern world is that the world is becoming more complex. Just as it takes a better trained person to understand and use a computer than to run and maintain an abacus, so it takes far more knowledge to really understand the interaction of events in an area like the Middle East or Central America today than in the past.

There is a ground rule, in college, that students should study two hours for every hour in class and that no professor should teach more than fifteen hours a week because professors need time to keep current in their fields.

A similar rule of thumb is necessary for the news media. Again and again, I am covered by reporters who have literally no idea what I am talking about. While Washington and New York have some specialists who spend time on defense policies, there are almost no specialists outside those cities who can write knowledgeably about the systems which may decide whether we can survive militarily and virtually no regional television stations in America with staff members who are serious students of Central America or the Middle East or any other topic.

Ignorance drives out knowledge in both political debate and news-media coverage. The less you know about a topic, the more likely you are to ask generic questions or to revert to folk wisdom. For most reporters, every conflict situation involving the United States raises the issue of "another Vietnam?" Never mind the reality that a Grenada or an El Salvador is far too small and too easily isolated by our forces to be a Vietnam: reporters ask that question because they don't know what to ask.

Similarly, the problems of military weapons procurement are treated as hunts for waste and scandal. This muckraking tradition is of some value, but thirty years of reporting on absurd purchases has done little to improve the process of procurement.

The search for failure and waste in Pentagon procurement is particularly damaging when reporters get test results on new equipment. One of the things young reporters are shocked to learn is that new equipment tends to fail. By definition, if you try to develop really new equipment, you are doing things you have never done before; therefore, you may not do them very well at first: if hardware is genuinely new, it has trouble-prone design aspects that remain unknown until field tests reveal them—as simulations cannot—to be in need of additional refinements.

This cycle of frustration and confusion is inevitable and unavoidable if we are to develop revolutionary weapons systems, a fact of life almost totally misunderstood by both the news media and elected politicians.

The inquest upon M-1 tank treads is a classic example of Congressional and news-media overkill: the army was trying to build the most advanced tank in the world; its designers consciously ran a number of risks to combine a new engine, new electronics, new armor, and new concepts to produce a tank that would be a generation ahead of any other.

The M-1 tank is, by all reports, an extraordinary vehicle. It has some weaknesses—it uses too much fuel—but it is nevertheless likely to dominate any battlefield against any other tank design. However, one of its teething problems was tank treads that wore out too quickly.

Critics of the tank, in and out of the Pentagon, took an enormous amount of time determining whether or not we should build this new tank because of these temporary problems with its treads. We know how to build treads; we have been building tank treads for sixty years; the problem was soluble. Yet the nit-picking, scandal-mongering sensation-seekers came very close to delaying the introduction (and thereby increasing the cost) of a desperately needed new weapons system.

What makes this process even more frustrating is that generally, even if a mistake is correctly highlighted, neither the news media nor the elected officials are sophisticated enough to correct the system for the future.

Effective management requires changing the systems of behavior and the structure of bureaucracy when they fail. That is the only way the legislative branch can really have an impact on the immense development programs of the Defense Department. Today, neither the elected officials nor the news media that covers them are competent to suggest systemic changes that cut to the heart of problems.

A new generation of reporters and politicians must work much harder at mastering the art of self-government and the art of survival. We must dedicate a great deal more time and effort to systematically considering the requirements of American survival, to learning the basic rules of war and of conflict management. Finally, we must develop far more sophisticated methods of running large bureaucracies and of explaining to the public what is happening in those bureaucracies.

This is a major challenge to both our reporters and our elected officials. Just learning where Grenada is, or how to pronounce Gemayel's name, is not going to be enough if America is to survive.

The news media are the nervous system of a free society. Our elected officials are the decision-makers responsible for our fate. No organism can survive if its nervous system is erratic and its decision-making centers are ignorant. A free nation needs hard-working and professional news media and competent, educated elected leadership.

This is the bedrock on which we must build our foreign and defense policies.

CHAPTER 11
STAR WARS AND SURVIVAL

The United States must learn to cope with certain new realities. First, we now live in an age of potential mass death in which nuclear and chemical-biological weapons may kill millions. Second, the speed and irretrievability of ballistic missiles puts the superpowers in a hair-trigger situation in which either might be forced to use its forces, or lose them in a sudden surprise assault. Third, the United States and the Soviet Union, far from being mirror images of each other, are in fact diametrically opposed both in ideology and culture.

These factors multiply to make the world more unstable and more dangerous. With each passing year the potential for destruction grows, the warning-time shrinks, and the gap between Soviet and American strategic policies widens.

Ballistic missile submarines patrolling close to the American coast have shortened our warning drastically, probably to under 7.5 minutes, a response time so short that it invites errors which could annihilate our two nations.

The differences between the Soviet and American military systems and the doctrines on which they are based create an inherently unstable situation. Soviet doctrine calls for targeting our command and control systems and training for a pre-emptive strike to eliminate our retaliatory capability. Soviet srategists argue that after an initial nuclear exchange the Soviet position will be superior to America's because their launch silos can be reloaded with additional intercontinental missiles. These facts suggest that the Soviets expect deterrence to fail and that they are grimly determined to maximize their chances of survival.

The Soviets are determined to be able to survive a nuclear conflict through a pre-emptive strike at the U.S. while their own civil defense program gives them the ability to withstand U.S. retaliation. Their intention is to either defeat our deterrent capacity or render it ineffective.

The Soviets, unlike most American experts, are students of Clausewitz and believe that "war is an extension of politics by other means." They believe it is ultimately possible to develop in their terms "a positive correlation of forces" which will give them an advantage in a crisis because they will be able to either bluff or eliminate our deterrence. In a crisis in which our deterrence is incapable of deterring, we might in effect have to accept a Soviet victory since annihilation would be the only alternative.

American liberals, on the other hand, have suggested than an inherently decent Soviet state has been threatened by an aggressive Western capitalism so that the Soviets simply have to protect themselves. In its radical Democratic form this becomes a constant apology for whatever the Soviets have done lately.

The Soviets, in fact, are taking advantage of our ideological and cultural differences by preparing for a showdown while encouraging our liberals to avoid confrontation—even at the expense of weakening our position. As a result, we become less able to think seriously about survival in the nuclear age while the Soviets become more so. President Carter's Secretary of Defense, Harold Brown, explained the current situation when he said, "When we build, they build; when we stop building, they build."

When the nuclear arms race began, the United States had an enormous lead. Gradually the Soviets reached approximate parity. For the last decade the free world has been stalled, trying to negotiate without building new weapons while the Soviets have been building as fast as they could. Throughout this period of "disarmament" negotiations, the window of vulnerability through which the Soviets could strip many of our nuclear deterrent forces in a pre-emptive strike has opened wider.

An American President who launched our nuclear arsenal on warning would be acting immorally and against the total American tradition.

However, if an American President waited long enough to verify that a real attack was in progress, he might find his nuclear forces so weakened that it would be madness to retaliate. If the Soviets were in the process of reloading their ICBM silos and still had most of their theater-level weapons available in Europe, most of their bombers available for one-way missions against the U.S., and most of their submarine-launched ballistic missiles intact, an American President would face the bitter reality that losing our missile fields was a human as well as a military disaster, that our great population centers, still untouched, were in effect hostages to Soviet forces.

In that setting, an American President might decide that negotiating some accommodation with the Soviet Union was the only realistic course left. Thus Soviet hegemony would have been established, the free world would have been dismantled, and the United States would survive only on Soviet sufferance.

How likely is that scenario? Far more likely than it was ten years ago.

The citizens of Western democracies have always realized that nuclear weapons are a threat to humanity. People are now becoming increasingly aware that the principle of Mutual Assured Destruction is a policy of bluff through mutual suicide notes. It is little wonder that the public is morally repulsed by a policy to which the acronym MAD is so appropriate. Mutual Assured Destruction is a bankrupt game-theory doctrine which uses civilian populations as hostages. Its rejection by ordinary citizens shows a better grasp of reality than do the sophisticated rationalizations of our national elite.

The problem with the nuclear freeze movement and similarly-focused opponents of Mutual Assured

Destruction is not with their motivation but with their strategy.

A nuclear freeze by the Western democracies would work if our competitor were another "rational" Western culture. Nuclear war is so clearly horrible that any reasonable society would be eager to negotiate an understanding.

The unilateral disarmament posture of the British Labor Party in the 1930s closely parallels that of the nuclear freeze advocates of today. The world learned that a policy of moral leadership through unilateral weakness does not work against a totalitarian regime. Wolves do not regard vulnerability by sheep as moral leadership but rather as the opportunity for a feast. Similarly, Jimmy Carter's policies of "moral leadership" and military weakness led to the Iranian hostage crisis, the Soviet invasion of Afghanistan, increased Communist aggression in Central America and the Caribbean, and expanded Soviet-Cuban influence in Africa. The nuclear freeze position becomes an absolute failure whether the Soviets disagree and the freeze is unilateral *or* if the Soviets agree and both sides freeze their weapons: in the first instance, the Soviets continue to grow in strength relative to the West; in the second, both the Soviets and the Americans have frozen themselves into Mutual Assured Destruction.

Given that MAD is an irrational doctrine and the freeze will not help, let us explore other options.

American defense policy should stem from the realization that both we and the Soviets will seek advantage through research and development efforts. Each side will feel itself constrained to at least "keep up" with the competitor to insure that the other side does not make a sudden breakthrough which changes the whole balance of power. Because this competition may well go on indefinitely, we should decide how we can minimize the chance that it will get out of control.

There are seven imperatives that will help the free world survive in the age of nuclear weapons while dealing with an aggressive and irreconcilable competitor.

1. Eliminate or minimize the chance of a pre-emptive strike by either side;

2. Lenghten the warning time to eliminate launch on warning as an operational alternative for either side;

3. Create layers of choice so that a mass destruction of population centers is not the only response available to an American President;

4. Improve Free World civil defense so that no Soviet planner can assume total victory from a nuclear strike;

5. Be capable of prosecuting a war effectively enough that Soviet military analysts will know that we are capable of surviving a first strike;

6. Prepare specific steps to meet the challenge of nuclear blackmail by countries with limited nuclear arsenals;

7. Recognize that, in a real conflict, undefended command, control, communications and intelligence satellites will inevitably be attacked, so that we must have redundant systems and antisatellite weapons of our own.

President Reagan's so-called "Star Wars" speech has to be seen in the context of Soviet doctrine, the horrors of Mutual Assured Destruction as a policy, and the need for a new approach to surmount the perils which threaten us.

The essence of the Reagan proposal is that the United States take advantage of emerging technologies to shift our focus from offensive weapons to defensive weapons. Advances in computers, in directed energy weapons, and in space technology open up new vistas for American survival.

This approach will deter a first strike by making its success uncertain. If the United States has no anti-missile defense, then Soviet planners can predict a successful first strike with near certainty if they simply apply two or three warheads to every American target.

The advocates of MAD have recognized that a successful preemptive strike is becoming possible and tried to find solutions through superhardening silos and by making large missiles transportable. Super hardening is ultimately pointless because a large Soviet warhead would leave the silo unusable, lying on its side in a crater, even if it survives the blast. Racetrack, mountain mining, and similar passive solutions have also broken down under analysis because they would be more expensive to develop than the countermeasures to overcome them.

We must undertake two initiatives. First, we must shift our retaliatory capacity to small, truck-borne missiles which can be driven all over U.S. military reservations. These "midgetmen" will create enormous targeting problems for Soviet planners, because it is almost impossible to insure their total destruction in a first strike.

Second, we must develop an active space defense so that a preemptive strike ceases to be a reliable instrument of Soviet power. This defense must include several layers of spaceborne instruments to observe the Soviet Union and to destroy missiles before they reach the United States—particularly in the boost phase, when ICBMs are most vulnerable and when they still contain all their warheads. A single Soviet SS-18 destroyed in the boost phase costs the Soviet planner from eight (the current set) to as many as twenty or thirty warheads (the estimated potential payload).

A layered antiballistic missile defense capable of significantly degrading a Soviet first strike can now be built, but critics argue that no such defense *should* be built because no defense could have a 100 percent effective against enemy missiles. But a "leakproof" defense, while desirable, is not necessary to our goal of forcing Soviet planners to abandon a preemptive strike because of uncertainty.

Liberals who warn us about the horrors of a nuclear war are nonetheless among the most active opponents of this system though it would make nuclear war less likely. Furthermore, their inconsistency is heightened when they point to cost as a major objection to an antimissile defense: the untold misery and destruction of a nuclear war would make even a partially effective missile defense America's best investment.

Because we still are in the early stages of developing the technology of a layered defense, we today can see only a general picture of how such a system might work. The system as now envisaged would include the following elements.

In addition to the current generation of reconnaissance satellites which constantly monitor the Soviet Union there would be mobile satellites surveying suspected Soviet submarine locations. As soon as the reconnaissance satellites picked up radio, infrared, and/or visual indicators of a hostile threat they would activate the command and control satellites which would be the brains of the initial defense.

A Soviet preemptive strike would by definition involve a large number of simultaneous missile launchings whose preperations provide massive amounts of warning data. The few missiles which the Soviets could prepose and launch under cover of legitimate activities (space launches, testing, etc.,) could be handled with relative ease by the point defense which I will describe below.

In a full preemptive strike, the command and control satellites would activate conventional, laser, and particle-beam weaponry to destroy the Soviet missiles.

Particle beam and laser solutions (directed energy weapons) offer some real advantages over conventional—gun and missile—weapons because directed energy weapons hit their targets at the speed of light, making aiming much easier and permitting a single weapon to hit multiple targets in series.

Once the surviving Soviet missiles leave the boost phase, they are vulnerable to attack during the transatmospheric portion of their flight. Since this requires a different sector, targets at these higher altitudes will probably require a second layer of defensive satellites. It is also conceivable that in some cases solid-fuel, rapid-launching rockets fired from the air, ground, or from ships could be used to further erode and weaken the Soviet assault in this phase.

About 20 or 30 miles downrange of our cities and silos, a point defense system, including lasers, particle beams, and high-acceleration rockets would engage.

It is conceivable that the most effective point defense will involve either satellite-based directed energy weapons or an orbital mirror system reflecting ground-based lasers. Such a system would be able to hit the less resistant rear of the reentry vehicle.

Critics of the "Star Wars" defense rely essentially on these arguments: first, that since a perfect defense cannot be built it is pointless to build any defense at all; second, that the proposed system is not technically feasible; third, that even if the system could be built it would be relatively easy for the Soviets to defeat it.

These arguments are precisely the kind of technically uninformed, defeatist thinking which led to the Allied conclusion that Czechoslovakia was indefensible in 1938.

We do not live in a perfect world of perfect solutions—we live in a relative world of problems which can be mitigated if they cannot be solved. If we can create a defensive system so daunting that the Soviets would fear to launch a preemptive strike, that breakthrough is worth almost any investment.

And yet the very people who claim we are discussing the fate of the earth suggest that an expense smaller than one year's payment on the interest on the national debt is too great an investment to make in our national survival.

Unrealistic demands for perfection lead to despair, inaction, and potential defeat or destruction. It is better to fund a research and development program for a layered space-defense system than to remain ignorant of its possibilities. At worst we will greatly increase the uncertainty factor in Soviet planning and therefore make a Soviet pre-emptive strike very unlikely.

The criterion by which we should judge strategic doctrines and weapons systems has to be one of a better or worse future rather than a perfect or hopeless future. Measured by that more realistic standard, there are only three broad strategic options for the next generation:

1. We could stay with Mutual Assured Destruction and watch public support continue to erode as people come to understand how much this MAD doctrine ensures their destruction if the bluff of deterrence fails. This approach will require continued emphasis on offensive weapons and a war-avoidance doctrine very susceptible to Soviet manipulation.

2. We could freeze development of new offensive or defensive weapons and pin all of our hopes on negotiations. We could lead by moral example and hope the Soviets and the terrorist nations will decide that moral courage without the force to protect it is worthy of respect. Given the record of the Soviet Union, Iran, Libya, Syria, and Cuba the notion that they care about moral strength is a bit far-fetched.

Attempts to provide moral examples through unilateral weakness have an even greater likelihood of failure than does the doctrine of Mutual Assured Destruction. There is at least a *hope* that the MAD doctrine may prevent war by its sheer horror; there is *no* hope that the U.S. could survive in a world in which we are weak and the forces of terror and totalitarianism are powerful.

3. We can go ahead with the Reagan initiative to develop a "star wars" active defense against ballistic missiles as a first step toward breaking out of the twin failures of MAD and the Freeze initiatives.

There are six realistic goals which would increase our children's chances of living in a world without nuclear war.

* * *

A. Move to less vulnerable, more mobile, and less visible offensive weapons such as Midgetman rather than MX;

B. Shift away from ballistic missiles toward recallable systems. Our bias should be toward bombers and super-quiet mini-submarines with cruise missiles which are more mistake tolerant and therefore safer in a crisis.

C. Continue to develop our intelligence-gathering capability in space and on earth. The more we know about another country, the longer we can wait in a crisis without accepting an unreasonable risk to our own security in pursuit of peace. Conversely, the more ignorant we are the more likely any given crisis will turn to violence. This principle requires that we will have to be prepared for some conflict in space because it is unrealistic to think that a serious superpower conflict could develop on earth in which our enemy would voluntarily let us retain inanimate assets giving us enormous command and control advantages in war-fighting. Soviet anti-satellite weapons require 1) that we develop redundancy in available systems so that we can replace lost satellites; and 2) that we have an anti-satellite system capable of keeping their assets hostage to deter them from attacking ours.

D. Put in place the active layered defense as described above. It will stop accidental launchings before people are killed; and, at least for the next half century, keep countries with limited technology and limited weapons in check. Further, by rendering the success of a Soviet pre-emptive strike uncertain, it will make the world far safer from nuclear war.

Possibly the most important contribution of a layered anti-missile defense would be that we would no longer need to consider launching our missiles on warning so we do not lose them in a pre-emptive strike. Those critics who claim that the space-based defense is vulnerable to Soviet hydrogen bombs miss the point: since the Soviet planner would know that any attack on the active defense would have to precede a pre-emptive strike, and any attack on the space-based defenses would guarantee the failure of such a strike by alerting us to its imminence. If a star wars active defense would simply eliminate the pressures of a Launch-On-Warning, that alone would be a great achievement worth the investment.

E. Move toward serious civil defense. While civil defense is of little value in a true spasm war, this is the least likely kind of nuclear war. In the three most likely uses of nuclear weapons—first by terrorists, second in a single warhead "demonstration effect" for diplomatic bargaining purposes, and third in a limited exchange involving only a few targets—the difference in human survival between real civil defense and the current pitiful effort would be enormous.

F. Strenghten our conventional forces to push back the escalation ladder. The free world must be prepared to win a conventional war. Today we could not. Today in Western Europe we would have to accept defeat in a conventional war or escalate to nuclear weapons within a relatively short time. That is not good enough. As a nation committed to civilization we must be willing to pay the price of being militarily strong enough to protect freedom without first resorting to weapons of mass destruction. It is possible with advanced technology and modern strategic doctrines for the West to defend itself solely with conventional forces. We must prove our moral commitment by committing the resources to be both free and moral.

The above six steps give us the potential to leave behind the horrifying dangers of Mutual Assured Destruction, the potential to defend ourselves with weapons alone; the potential to protect our citizens from even the most horrifying efforts of terrorists.

For too long we have measured our survival against our immediate desires. In the long run that is a recipe for disaster: we must measure defense requirements, not defense fantasies. If we would give to our children and grandchildren a safe America that is free and peaceful, we must learn the hard lessons of a forty-year competition with the Soviet Union, and consider the growth of terrorism and terror nations. Only then can we accurately assess our real defense requirements and build an America capable of surviving.

CHAPTER 12
MANAGING STAR WARS

The most important change in war-fighting, national defensive strategy, and foreign policy since the advent of the A-bomb and counterforce is the emerging potential to protect both freedom and lives through the development of space.

Through proper space management and the creation of space-based defenses of consequence, we have an unparalleled opportunity to free our children and grandchildren from the spector of nuclear war—forever. Space-based laser and stripped particle beam systems, coupled with conventional weaponry in low orbit and mounted on spaceplanes, and augmented by ground-based antiballistic missiles, could eventually make nuclear war impractical.

The Soviet Union already has tested killer satellites. Intelligence sources speculate that, in violation of the treaties banning "weapons of mass destruction in space," the Soviets have nuclear weapons in orbit. On the ground, they certainly have placed an antiballistic missile defense system around Moscow. The Soviets are also testing an air defense/ballistic missile defense weapon and have placed missile defense radars at strategic Soviet sites north of Moscow and at Abalakova in what may be a violation of the mutually-agreed-upon Antiballistic Missile Treaty.

A White House official remarked in an article in the January 16, 1984, issue of *Aviation Week & Space Technology* that "The CIA's position on Soviet ballistic missile defense activities is unusual in its strength and clarity—an alarm bell that we must watch the situation very closely."

Thus, this "Star Wars" era is not one we enter by

choice. The creation of a National Space Command is no longer optional since the Soviets are pouring massive resources into space defense and its corollary, ballistic missle defense. And yet, in typical welfare-state style, even before the nation is truly committed to protecting its interests in space, lobbying groups are gathering to inject into our defensive space effort every element which has made our government so ineffective in managing policy and procurement issues here on Earth.

While knowledgeable officials in Washington are concerned over a lack of "national technical means" —photoreconnaissance satellites such as those used to monitor the Soviet Union's compliance in arms control agreements—and the Soviets develop and deploy their camouflage, concealment and deception program to counter our satellites' capabilities, Liberals block funding on research and development programs so that we have had to stop producing current types of reconnaissance satellites in order to develop replacements—which may turn out to be unsatisfactory.

While the glamour issues—laser battle stations and weapons—receive all the media attention, government agencies are scrambling for their piece of pie-in-space, offering competing design studies and bureaucratic management teams. Instead of a unified Space Command, we are told we need multiple space commands—separate groups monitoring different projects from the Air Force, Army, Defense Advanced Research Projects Agency, and many others. If our defensive space effort is not to go the way of the Rapid Deployment Force, we must halt this proliferation of bureaucratic layering before it becomes entrenched.

For example, in the area of spaceplane design, a number of studies have been done on transatmospheric vehicles capable of attaining Mach 29 and taking high-resolution photographs of sites on Earth from 45 miles up; but Air Force sources estimate that such craft could cost $100 million apiece, so they fear to request serious development money.

Meanwhile, the Defense Advanced Research Projects Agency has asked for $11 million to study a manned space cruiser which could perform military manned space missions. This much flashier space vehicle would theoretically be capable of research and development missions, hypersonic flight, and even rescue missions to the Moon—but it must be piggy-backed into orbit on the shuttle, and it will be piloted by a single man in an open cockpit, protected only by an anti-micrometeorite suit supplied by the lowest bidder. Although such a multipurpose space vehicle has uses, a transatmospheric vehicle which could replace the jets of today in the world of tomorrow is crucial. The spaceplane that we build, however, will be chosen by the military bureaucracies and Congress with an eye to political advantage, not what will be advantageous for the country.

There is a great danger that the decisions to fund or not to fund various components of our space effort will be made on a basis of patronage, politics and pork-barrel, rather than on the nation's need for survival. We need *both* the station and antimissile defenses, and yet the bureaucracy systematically undermines Presidential initiatives. Where Reagan offers bold strokes, they pursue bureaucracy as usual.

Critics, among them some defense analysts, who feel that ballistic missile defense would increase East/West tensions, or that laser weapons may not be effective, add to the Liberal knee-jerk reaction that *any* high-tech weapon is an escalation of the threat of mass annihilation.

Quietly, among experts, a consensus is growing that a defense-oriented world is better than an offense-oriented world in which the doctrine of Mutual Assured Destruction continues to prevail. Mutual Assured Destruction (MAD) has had results which those who crafted the doctrine surely did not anticipate: it has escalated the arms race, terrified the world's populace, and made strong governments appear to be juggernauts of destruction. America has been indicted and convicted simply because it has protected its citizens in an ever more dangerous world. This image, in turn, has alienated succeeding generations of young people from science, from technology, from the military and from government.

Young people tend to be idealistic. Their idealism is manipulated cannily by Soviet propagandists to make it seem that the West alone is responsible for the arms race, when the doctrinal insanity of MAD—mutually agreed upon by the superpowers—is at fault. Among other things, MAD has given us detente, a framework whose greatest success has been to extract from the peace-seeking West better than a trillion dollars in loans to Communist Bloc nations. It is well known but seldom mentioned that the nuclear missiles the Soviets targeted at NATO were paid for by detente-era loans from Western governments eager to "improve relations with our communist neighbors." This welfare-state approach to geopolitics has demonstrably worsened relations worldwide.

There are alternatives to Mutual Assured Destruction other than unilateral disarmament or buildown—there is the doctrine of Mutual Assured Survival based on a defensive capability being attained by both the U.S. *and* the U.S.S.R; there is Defense Protected Buildown, a doctrine which would link the development of defensive weapons to a reduction in offensive weapons, thus allaying fears that a new round of U.S.-Soviet arms competition will lead to new weapons systems capable of penetrating any such defense.

An opportunity-society approach to the question of which path to follow must consider the humanity-conserving aspects of each system.

MAD has proved to be dehumanizing, counterproductive, and costly; Mutual Assured Survival, by itself, assures little because it addresses only additional layers of weaponry, not reductions in the nuclear arsenal that has haunted the dreams of our young for too many generations.

Defense Protected Buildown, bringing with it a reduction in nuclear weapons as well as the technological breakthroughs in laser and particle beam technology and space reconnaissance, offers the greatest hope for a safe future. And a safe future is what we want for our grandchildren.

No technology will make nuclear weapons obsolete by itself. No defensive system can be built that will destroy all incoming missiles. The best that can be done is to make further production of nuclear weapons unattractive because those weapons have a low probability of hitting their targets, because the psychological-warfare benefits of building them and deploying them have decreased, and because if war-preparation is effective, war-fighting need not occur. There is an adage, among the military, that an army has failed once it must fight. An army's main purpose is to deter aggression, not meet it.

Liberals have forgotten this, and thus bring us closer to war by their continual demands for peace at any price. The welfare-state factions that cry out for unilateral disarmament are actually inviting confrontation. The welfare-state politicians who oppose research and development appropriations on the grounds that this money would be better spent on social programs fail to understand that most of the high technology advances seen in the last 40 years have come from the military and intelligence communities.

Research and development money is crucial to our economy because it demonstrates the Federal government's commitment to specific areas of endeavor and encourages independent research and development by high-tech companies. Today, new high-tech companies are proliferating, especially in aerospace, but those companies are barred from bidding on many government projects by complex procurement procedures put in place by special interests. Even the larger companies involved in laser and particle beam research are waiting for some indication that we *are* serious about a multi-layered defense: the $1.3 billion allotted for 1985 is little more than was allotted to previous development, and is insufficient to excite large in-house programs by contractors not already assured of a piece of the pie-in-space.

At this writing, there are over a dozen agencies

involved in managing the development of our space defense system. What we have already learned about welfare-state protocol proves that when so many interests are involved in a project, that project becomes hideously expensive, inefficient, and self-sustaining.

Given these realities, an opportunity-society approach would be to open the contest for research and development money more widely, form a unified Space Defense Command not beholden to special interests, and institute a program whereby America reduces its stockpile of offensive weapons as a ballistic missile defense is put in place—a move which would assure the Soviets that we are not escalating the arms race but seeking a new and safer parity for the citizens of the world.

CONCLUSION: WHAT CAN EACH OF US DO?

If *Window of Opportunity* has any meaning it is as a blueprint for a very different, more optimistic, decentralized, growth-oriented, safer American future. This future will not be achieved by a handful of politicians in Washington dictating to the rest of us.

The heart of the futurist thesis is that we are moving toward a more decentralized, participative society. The key notions in Peters and Waterman's *In Search of Excellence* all relate to arousing strong performances from individuals and from small teams. Effective large companies are the ones which behave as though they were collections of smaller organizations.

The need for decentralized, human-sized efforts makes this concluding chapter particularly important. The real key to an information-age opportunity society is true power to the people—not phony liberal-populist cries of power to the people which really mean power to the bureaucracy, and not phony big-business conservative cries of decentralized power in Washington when they secretly want centralized power in the board room and the country club.

The problem with a true commitment to power to the people is that it requires the people to exert themselves—as the Founding Fathers well understood.

Thus Thomas Paine wrote scornfully of sunshine patriots and summer soldiers, and Benjamin Franklin, when asked what the Constitutional Convention had achieved, replied, "A republic—if you can keep it."

For America to remain free, safe, and prosperous, we all must remember James Madison's statement: "Knowledge will forever govern ignorance: and a people who mean to be their own governors must arm themselves with the power which only knowledge gives."

The acquisition of knowledge is not an easy or instant process. For members of a generation that grew up on fast food and packaged entertainment, the process of arming themselves with knowledge seems particularly daunting. Yet is is the unavoidable first and most basic cost of the survival of freedom.

People are not in general stupid, but they are often ignorant. In their ignorance they often tolerate ignorant news reporters who in turn tolerate ignorant politicians. The result is an ignorant politician making an ignorant speech to be covered by an ignorant reporter and shown in a forty-second clip on television to an ignorant audience.

For a nation which has avoided learning the basic rules of history, of economics, of military and foreign policy, we have done remarkably well for half a century—but time is running out.

The generation that fought World War II and governed America into the 1960s dealt effectively with a changing world because it had studied issues and ideas in depth. Eisenhower, often laughed at as a non-intellectual by liberal academicians, spent part of his time as President of Columbia University in the late 1940s translating Thucydides from the original Greek to regain mastery of the language.

The generation which led America after World War II was made up of men who had studied extensively, traveled widely, and learned hard lessons through two world wars and a long decade of depression and appeasement in the 1930s. They had been schooled in what John F. Kennedy would call a "hard and bitter peace."

Our generation has decayed in learning, in realism, and in toughness from our fathers' generations. We throw temper tantrums on television about corrupt politicians but then go back to vacationing and gossiping without having done anything. We complain about waste in the Pentagon—but then fail to follow through with real reform; about our children's schooling—but refuse to invest the time and effort to improve it. We ignore opportunities to speak at public meetings, elect local officials, educate our legislators, demand professionalism of our news media, —and then wonder why the world isn't working very well.

If you are willing to spend the time and effort of creating a true window of opportunity as the greatest gift you could give your children and grandchildren then these are my recommendations to you.

PRINCIPLES OF SELF-GOVERNMENT FOR AN OPPORTUNITY SOCIETY

One of the problems undermining our whole system of government is the influence of Machiavelli on political science departments and through them on the news media. Machiavelli's cynical, corrupt view of manipulation and the use of power to govern through fear underlies much of our modern inability to talk about self-government.

We badly need an honest, practical, yet idealistic citizen's guide to the process of self-government as the leading nation of the Western Alliance. In the absence of such a guide I propose these principles of involvement.

1. Every American is ultimately responsible for a share of the American system. In a dictatorship people can legitimately stand to one side and wait for the dictator to succeed or fail. In a free society we each have an obligation to do our part to save our nation for our children and grandchildren. You do not have the moral right to wait for Reagan or Mondale or Gingrich or your local congressman to save the nation. You have an obligation to do your share to help your elected leaders save the nation, and a further obligation to try to oust them if in your judgment they are failing in their duties.

2. Your role as an effective citizen requires far more knowledge than most citizens realize: you have to know how your government works; you have to know what you believe in and how your beliefs re-

late to reality; and you have to know how your elected officials behave and whether they are effective in fighting for your beliefs.

3. As a citizen you need a mechanism to keep track of your elected official's promises and their actual behavior. The most powerful tool we have invented yet for this purpose is the Political Action Committee. PACs permit as few as fifty people to file with the Federal Election Commission. A PAC that contributes to at least five candidates can disburse up to $5,000 in an election cycle.

I strongly favor PACs because they tie candidates' promises into their performances by keeping records more effectively than do individuals. By linking their contributions to performance in areas of interest to the contributors, the PAC system encourages more people to be involved because it makes their contribution and their endorsement more effective.

The result will not be the control of candidates by any specific set of special interests but rather the ability of candidates to appeal to wide variety of PACs to find the ones with whom they share beliefs and goals.

This proliferation of open, publicly registered and publicly monitored support is in the best tradition of participatory democracy and is the first step for average Americans in learning to organize and systematize their new relationship with elected politicians.

4. If you are on cable television, you should insist that it carry C-SPAN—the television channel which covers the House of Representatives. The programs which Brian Lamb at C-Span has developed when the House is out of session included a brilliant coverage of one of the Iowa caucuses, a model of how America really tries to govern itself and how average individuals can make a real impact on their government. You should also write your senators and ask them when they are going to come out of the dark ages and begin to televise Senate proceedings.

5. You should occasionally go to your city, county and school board meetings. Even one visit a year to each will teach you a lot about what is going on in

your community. If you take your children along, they will begin to learn about your community and the process of self-government.

If your local governments aren't worth one meeting a year to you, then you might reconsider the responsibilities of a free citizen.

6. You should use your local public library. While television is a useful teaching device, it is too unfocused a medium for the volume of detailed knowledge a self-governing citizen needs. Your local library will provide you with manifold learning opportunities at little cost and it is in fact one of the key solutions to the link between poverty and ignorance. People may be born poor, but almost everyone in America is within walking distance of their library.

Remember that Benjamin Franklin helped found the first North American public library precisely because he associated knowledge with self-government.

7. You ought to found an opportunity society club in your community, neighborhood, or professional-occupational group.

By talking with local reporters, calling on the local radio talk show and having a few friends make announcements to local civic groups, you could probably find fifteen or twenty people interested in meeting once a month to discuss the future and its local implications.

It is in applying the general principles of an opportunity society to your town, your school system, and your occupation that will really begin to change the world. If we can get thousands of Americans to experiment with the future independently, then we will really build the momentum to lead the planet to a future of freedom, prosperity and peace.

8. Buy a home computer for you and your family. The computer is more than just a toy or a tool: it is a step toward a new range of possibilities which we will only understand by practice and by experiment.

The sooner your children can begin playing with the computers which will be the dominant instruments shaping their future, the better their chances of getting good jobs in that future.

Computer networks will, relatively inexpensively, allow you and your friends in local clubs to communicate by electronic mail.

Working both for pleasure and for profit, you and your friends can help invent the computerized neighborhoods, schools, and jobs of the future so that more and more people will have an outlet for their concerns and a source for effective information.

9. Finally, try, try again. Self-government is an arduous, demanding task on which the survival of freedom depends.

We stand at the pinnacle of man's long climb from savagery to a truly free civilization exploring our planet, our solar system and our universe. We can glimpse the future through our window of opportunity.

The adventure continues—be part of it.

POSTSCRIPT

This book is the product of many people working to develop a concept which some of us in Congress came to call the Conservative Opportunity Society. The original idea for developing a new positive vision and theme to replace the liberal welfare state as a model for America's future began in discussions with Paul Weyrich of the Free Congress Foundation and Ed Feulner of the Heritage Foundation. Richard Viguerie first gave us national attention and discussed new ideas in the *Conservative Digest*.

With their assistance in polling and research, Marianne and I wrote a series of papers which were critiqued by Pete Jensen of Georgia Tech, Steve Hanser and Mel Steely of West Georgia College, Daryl Connor of O.D. Resources Inc., Jim Tilton of Hunton and Williams. Congressman Guy Vander Jagt and the National Republican Congressional Campaign Committee—beginning with Steve Stockmayer, Wilma Goldstein, Russ Evans, Nancy Sinnott, and continued today by Joe Gaylord, Linda Divall and Ed Goeas—played a key role in developing these ideas.

Dicussions about public opinion and extensive polling were developed with Bob Teeter and Fred Steeper of Market Opinion Research and with Dick Wirthlin and Gary Lawrence at Decision Making Information. The vital role of the news media as the irreplaceable nervous system of a free society has been developed through long conversations over the years with Reg Murphy, Art Pine, Bob Novak, Andy Cassells, Jim Minter, and Dick Williams. As a critic, reporter, and thinker in his own right, David Broder has proven indispensable.

No two figures have played a greater role in the development of the whole Conservative Opportunity Society concept than Eddie Mahe and Bob Weed. Their long-range vision, sophistication, and commitment make them unique among political consultants. Jim Muncy, founder of Using Space for America Foundation, Diana Hoyt of Congressman Dan Akkaka's office, the driving force behind the Congressional Space Caucus, Ben Bova, Dr. Hans Mark of NASA, and Dr. Edward Teller played key roles in discussing space and high technology.

Alvin and Heidi Toffler have been good friends who often disagree with us about political philosophy but have broadened our vision and enriched both our work and our lives. Richard Rahn of the U.S. Chamber of Commerce, Burt Pines of the Heritage Foundation, John Mueller in Jack Kemp's office, and Jude Wanniski have all helped shape and the economic and political policy aspects of the movement. Dave Hoppe of Trent Lott's office, Don Eberly of the Republican Study Committee, Margo Carlisle of the Senate Republican Conference and Dick Thompson of the Senate Republican Policy Committee, and Morton Blackwell, President of Leadership Institute, have all been of great help in rethinking our effort.

The Republican Party has been invigorated in recent years by the neoconservative movement. Congressmen Mickey Edwards and Jack Kemp played key roles in bringing us together with Ambassador Jeane Kirkpatrick, the intellectual godfather Irving Kristol, and others. Peter Drucker put us on the right track in a conversation where he suggested that he had principles but no answers to the problems of our generation. He asserted we would have to make our own breakthroughs and mistakes for our generation, so we undertook just that endeavor. Dr. Ed Hall and his wife Mildy were informative and encouraging in several long conversations about the importance of simplicity, humanism, decentralization and rethinking the bureaucracy of the liberal welfare state. Frank Gregorsky has been my adviser in writing skills, my ally in research, and my critic in intellectual integrity as we worked on this new COS vision. David Warnick was our original writer and researcher in developing a model of self-government more oriented to traditional values and the neighborhood.

Many of my concepts of a holistic approach to self-government and a less mechanistic vision of the world came from teaching at West Georgia College in Carrollton, Georgia. When I arrived in 1970, West Georgia was a truly remarkable campus. It had the only humanistic psychology program in the East and a commitment to develop an "interdisciplinary futures environmental studies program," as well as an administration and faculty excited about ideas. In particular, I am grateful to Ward Pafford, John Martin, Chuck Masters (now with the U.S. Geological Survey), Cecil Fox, Dick Dangle, Ben Kennedy, Mike Arons, Jim Klee and Don Wells for the time they spent instructing me. While some of them would disagree strongly with my philosophy, all were exciting men to work with.

Ken Duberstein, former chief of Legislative Liaison at the Reagan White House, B. Oglesby, his successor, Bruce Chapman at the White House Long Range Planning Office, Lee Atwater, formerly in the White House and now part of the Reagan reelection campaign, and Dennis Thomas, formerly at the Treasury Department and now House Liaison at the White House, deserve special thanks for teaching me about the executive branch and helping us in a variety of ways.

Major General Don Morelli, U.S. Army (ret), the developer of the AirLand Battle, Colonel John Boyd, U.S. Air Force (ret), Colonel Harry Summers and Colonel Wally Franz at the Army War College, Dr. Tom Etzold at the Naval War College, Lieutenant Colonel Dave Drew of the Air Force University, Colonel Trevor Dupuy, U.S. Army (ret) and one of our leading students of the use of history, Dr. Forrest Pogue, Bill Lind of Senator Gary Hart's staff, Franklin Spinney at the Department of Defense, (possibly our most insightful planner), Jim Woolsey of the Center for Strategic and International Studies at

Georgetown (among other credentials), and Mike Burns, who spent several years developing the intellectual framework for our efforts on military reform to help shape the thinking about defense and foreign policy.

My Congressional and campaign staffs have played a tremendous role in freeing my time so that I could focus on thinking and research. Without their daily efforts to serve the 6th District of Georgia and respond instantly to each new crisis, it would have been impossible to serve in Congress and write a serious book.

The entire Conservative Opportunity Society effort would have been an intellectual concept instead of a serious political movement without the commitment and hard work of a number of Congressmen. In particular, the actual COS Wednesday Meeting Group, which hammered out many of the ideas, played a key role. Vin Weber, our Chairman, Ed Bethune, Dan Coats, Mickey Edwards, Judd Gregg, Duncan Hunter, Dan Lungren, Connie Mack, Mark Siljander, Barbara Vucanovich, and Bob Walker have been wonderful colleagues to work with. Finally, all of this enthusiasm would have gone for naught were it not for the efforts of our Coordinator, Janis Kerrigan-Roberts.

Other members of the House who have made a particular impact upon this effort are Dave Stockman, both during his House service and in his role as the Director of the Office of Management and Budget; Phil Gramm, on both budget and strategy matters; Bill Whitehurst and Jim Courter, in developing the Military Reform Caucus, just to name a few key players in what would otherwise be a very long list indeed. The COS group could not have prospered without the active suuport of the House GOP leadership. John Rhodes, the Minority Leader when I arrived, was remarkably patient and tolerant as we experimented. His successor, Bob Michel, participated in building our momentum. Trent Lott, Dick Cheney, Jim Martin, and Jerry Lewis have been tremendous. Jack Kemp is the creator of a dynamic intellectual attitude among House Republicans; his impact in creating an energetic, idea-oriented GOP may well make him the most important Republican since Theodore Roosevelt.

Finally, this book would have been literally impossible without the help of my wife, Marianne, and two great professional writers, David Drake and Janet Morris. Marianne has been a partner in developing the Opportunity Society concept from the very first paper, back in 1981. However, all her efforts would have been for naught without our two teachers and compatriots in the art of writing. David Drake is a delightful person, a good student of history, and a first-rate writer of adventures past, present, and future. His patient cheerful comradeship carried the book from Jim Baen's initial conversation to a working manuscript of respectable import. Janet Morris proved to be the teacher, taskmaster, and keeper of quality standards for all of us. Originally asked to polish our manuscript, she rose to the challenge and then immersed herself with David in a serious, thoughtful process of rethinking and rewriting which dramatically improved the book. As a writing team, David and Janet turned out to be a wonderful complement to the work Marianne and I did on the ideas and concepts of the first draft. We will always be grateful to them and we're in their debt.

While I am deeply indebted to everyone mentioned and many others who have helped shape my thinking and my skills, this book remains ultimately my responsibility. Whatever errors or mistakes occur in it are mine and I am to blame. Whatever felicitous insights and concepts emerge from these pages can be attributed to the many people who have shared the dream that freedom can spread, that prosperity can be available for all, and that peace can be obtained through hard work and tough-minded realism. As President Reagan taught us, we have every right to dream heroic dreams. That is the commitment which brought us together.